Specious Science

Specious Science

*How Genetics and Evolution
Reveal Why Medical Research on
Animals Harms Humans*

C. Ray Greek, MD, and
Jean Swingle Greek, DVM

continuum
NEW YORK • LONDON

2003

The Continuum International Publishing Group Inc
15 East 26th Street, New York, NY 10010

The Continuum International Publishing Group Ltd
The Tower Building, 11 York Road, London SE1 7NX

Printed in the United States of America

Library of Congress Cataloging-in-Publication Data

Greek, C. Ray.
    Specious science : how genetics and evolution reveal why medical research
on animals harms humans / C. Ray Greek and Jean Swingle Greek.
        p. cm.
    Includes bibliographical references and index.
    ISBN 0-8264-1398-6 (hardcover) 0-8264-1538-5 (paperback)
        1.   Diseases--Animal models. 2. Animal models in research.
    3.   Animal experimentation I. Greek,, Jean Swingle. II. Title.

    RB125 .G744 2002
    619---dc21

                                                    2001056191

We dedicate this book to all our patients

Contents

Acknowledgments

Many people and organizations have contributed to this volume. Although we have attempted to include everyone, we will undoubtedly leave some out and for that we do apologize. Thank you to Rick Bogle, Peggy Cunniff, Pamelyn Ferdin, Steve Fuller, Susan Green, Clare Haggarty, Dawn Haney, Larry Hansen, Nancy Harrison, Pandora Pound, Jill Russell, Jerry Vlasak, and C.W.

Kathy Archibald pointed out errors both large and small and made the book much better than it would have been without her editing.

We gratefully acknowledge that the following groups provided material for us: The Medical Research Modernization Committee, the Physicians Committee for Responsible Medicine, the New England Anti-Vivisection Society, and the American Anti-Vivisection Society.

A special thanks to the National Anti-Vivisection Society for whom Ray is the Scientific Advisor and Jean a veterinary consultant. Whenever we are asked for resources about the ethical arguments against animal experimentation, we refer people to NAVS's excellent web site, which can be found at *www.navs.org*. NAVS provided us with resources, without which this book could not have been written.

A special thanks to Mark for his invaluable critiques.

A special thanks to Tershia D'Elgin and Rita Vander Meulen for their skills and abilities.

And finally we want to thank our agent Julie Castiglia and our editor Evander Lomke who continue to offer great advice and assistance.

Preface

This book is a scientific examination of the ineffectiveness of using animals as models for human disease. It contains data and concepts suited to a college-level course in biology. We know this book will challenge some of you who do not have a science background (it may challenge some of you who do), but we hope you accept the challenge this book offers because a thorough understanding of science, the scientific method, and why animal models fail to cure human disease are concepts of vital importance to you, your family, and loved ones. In the following chapters, we explore our assertion that, because animal studies are not predictive of what will occur in humans, the extrapolation of results from animal models misleads scientists and harms human patients. Animal-based research therefore wastes time, money, and talent— all invaluable resources that can be put to use in far more effective ways for the betterment of human health and safety.

At the same time, this is *not* a book about animal rights. Although we certainly have strong opinions on the philosophical and ethical debate regarding animal rights and other social issues, such as capital punishment and abortion, in *Specious Science* we concern ourselves exclusively with science. As scientists and doctors, we stand as rigorous defenders of science, rational thought, and the trial-and-error process of scientific discovery. That is not to say we believe that science is the only way to know reality. There are others, such as intuition, religion or spirituality, and aesthetics. However, we—and those who experiment on animals— believe that science is the chief means of understanding the reality of our *material* world, and specifically the reality that is biomedical science.

As you will see, our case against the animal model, as we present it in *Specious Science,* and as we previously presented it in our first book, *Sacred Cows and Golden Geese: The Human Cost of Experiments on Animals,* rests entirely on the fact that animals are inappropriate models for human disease. As much as we stand by our assertion that science *does* work, we stand equally behind our assertion that the paradigm that scientists who uses animals as a models for human disease rely on absolutely does *not*. In that sense, we consider animal models to be as much a pseudoscience as extrasensory perception, phrenology, astrology, and the various forms of "alternative" medicine, such as magnet therapy, energy fields, qi, Lakota medicine wheels, laying on of hands, homeopathy, and the like.

In *Specious Science,* we hope to educate (or remind) you, the reader, about what true science is and about research that sounds like science but isn't. You'll also discover what happens when science as a process comes into conflict with scientists who happen to be human beings with egos, mortgages, and children to put through college.

In the first two chapters, we examine how using animal models to study human disease violates the philosophical underpinnings of science itself. In discussing why animal models for the study of human disease and treatment are not an application of science, we set forth the scientific basis of why experiments on animals do not help humans. These chapters introduce the concept of science and why an understanding of the difference between science and pseudoscience is vital if we are to understand why experiments on one species cannot lead to reliable data about another.

In the remaining chapters, we explore how medical advancements are really made and specific cases of how the extrapolation of results from animal models has misled scientists and harmed human patients both directly and indirectly. The direct harm, as you will see, lies in subjecting humans to damaging treatments simply because they were shown to be safe and effective in animals. Indirectly, animal data delays life-saving discoveries and wastes valuable resources. Throughout these chapters, you will find a wealth of examples of how animal models give misleading, wrong, and dangerous results when applied to humans. We also address how many of the discoveries from animal models could have occurred without using animals. (Animals are subject to the same laws of physics, chemistry, and biology as humans, but were not *necessary* to prove such laws.)

It is our hope that after reading *Specious Science,* you will come away with a thorough, fact-based understanding of why experiments on animals waste time, money, and talent. Reading the entire book chapter by chapter is the best way to accomplish this. However, we understand that some readers may fear that the chapters one and two may pose too great a challenge for those not well-versed in science. If you feel that way, be assured that skipping them will not prevent you from gaining valuable information from the rest of the book. At the same time, we strongly urge you to give chapters one and two a try. Having a firm grasp of the philosophy of science and the scientific method will help you better understand the debate about animal models, and we have made every effort to ensure that the material is as "reader friendly" as possible. Moreover, we believe that it's important for people to take responsibility and learn the basics of science as it affects their daily lives. Reading chapters one and two—indeed, *Specious Science* as a whole—is a good opportunity to do so.

The public should know that there are a number of individuals and groups in the scientific community who, despite specific knowledge of the failure of animal models and their role in harming human patients, maintain a financial interest in perpetuating the practice. These groups include scientists, pharmaceutical companies, healthcare providers, manufacturers of animal-laboratory equipment, and even scientific journals. They believe that they are entitled to police themselves; that is, they wish for nothing less than total control over the allocation of the public funds spent on biomedical research. They insist that the average American cannot understand science, and therefore should not have an opinion as to how taxpayer dollars should be allocated for research.

We disagree. Since the publication of our first book, *Sacred Cows and Golden Geese*, we have been criss-crossing North America, Europe, and Australia, delivering hundreds of lectures on the subject of using animal models to study human disease. In our travels, we have had the privilege of meeting people from all walks of life who have no difficulty understanding the concepts we present, and they tell us that what we are saying makes sense. Whether they were Ph.D.s with diverse life experiences or folks who have never left the neighborhood where they were born, they all shared one trait: they knew that money talks. No one in any of our audiences needed in-depth knowledge of the theory of relativity to understand that putting vested-interest groups in charge of allocating research funds is akin to putting the fox in charge of the hen house.

Despite our grass-roots appeal, it has been difficult to get our articles published in newspapers and magazines and to get airtime on television. Because newspaper and magazine editors, as well as television executives, depend on advertising revenues for their sustenance, they are loathe to offend potential "big money" advertisers, such as pharmaceutical companies and health-care providers, by presenting views that may cast them in an unflattering light. *Specious Science* is an opportunity to reach the public with an authoritative, well-documented examination of why using animals as models for humans in biomedical research fails as a scientific paradigm and how the pseudoscience of animal experimentation harms human patients and delays life-saving discoveries. You will find here an examination based purely on sound scientific theory and principles, devoid of the kind of rhetoric and emotion that, regardless of its virtue, has no place in such discussions. As emissaries from the land of science, we are not here to expound on the ethics of using one species for the benefit of another. Our goal is to examine the animal model purely from a scientific viewpoint and reveal it for what it is—an unequivocal failure that works to the detriment, not the benefit, of human health.

chapter 1

The Philosophy of Science

Ever since puberty, I have believed in the value of two things:
kindness and clear thinking. At first these two remained more
or less distinct; when I felt triumphant I believed most in clear
thinking, and in the opposite mood I believed most in
kindness. Gradually, the two have come more and more
together in my feelings. I find that much unclear thought
exists as an excuse for cruelty and that much cruelty is
prompted by superstitious beliefs.[1]

—Bertrand Russell (1872–1970), British philosopher
and mathematician

There are many ways of knowing *truth,* including intuition, religion,
and spirituality. However, it is generally agreed that science, which deals
with observed facts and the relationships among those facts, is the chief
means of understanding the reality of our *material* world. Thomas H.
Huxley, British zoologist and author advised those seeking truth to "Sit
down before fact as a little child, be prepared to give up every precon-
ceived notion, follow humbly wherever and to whatever abyss nature
leads, or you shall learn nothing."

The word *science* comes from the Latin word *scientia,* which means
knowledge. Throughout the ages, philosophers as well as practitioners
in the field have sought to provide a definition of science that captures
the goals and values of this broad field of endeavor. As easy as it is to
list the characteristics or underlying principles of science, a succinct
definition of science can seem elusive.

In his book *Consilience,* Harvard biologist E. O. Wilson writes "Science is the organized, systematic enterprise that gathers knowledge about
the world and condenses the knowledge into testable laws and principles." Stephen J. Gould defines science as "a teaching authority dedicated to using the mental methods and observational techniques
validated by success and experience as particularly well suited for de-

scribing, and attempting to explain, the factual construction of nature." Thomas H. Huxley stated "Science is simply common sense at its best; that is rigidly accurate in observation and merciless to fallacy in logic." In practice, science is the observation, identification, description, experimental investigation, and theoretical explanation of natural phenomena. It demands systematic methodology and study, as well as internal consistency. In the world of science, no assumption is allowed to go unchallenged.

How do we define pseudoscience? The prefix *pseudo* comes from the Greek word meaning *false*. So pseudoscience is literally *false science*. Moreover, it is false science that claims to be real. As opposed to true science, pseudoscience can be characterized by static or randomly changing ideas, the use of vague mechanisms to acquire understanding, the lack of rigorous logic and organized skepticism, and disregard of established results.

We all use the principles of genuine science—trial and error, common sense, and logic—in our everyday life. On our way to work in the morning, we turn down a certain street because we have learned through trial and error that it is the fastest route to the office. At lunch, common sense tells us to pass on the fish special, since last week it caused gastrointestinal distress. In the evening, it's only logical to turn on our favorite TV program because we know from experience how much we enjoy it. This may seem very obvious, but the alternative would be chaos—constantly selecting products at random, with absolutely no continuity. From anthropology to zoology, the world of science encompasses numerous fields of study. Whether an anthropologist is excavating the remains of a prehistoric civilization or a zoologist is observing the mating habits of the lowland gorilla, these scientists are studying the workings of the world and the relationships among these workings.

The goal of science is to discover the truth. Science searches out the answers to questions and solutions to problems. It also attempts to satisfy curiosity. Although it seeks to gather more facts about the world we live in, science involves much more than the accumulation of facts. Martin Curd and J. A. Cover state:

> ... Truth by itself cannot be sufficient as a characteristic of the goal of science. This is relevant because so many of the true statements we could make about the natural world have little or no scientific value. Imagine, for example, that a biologist wants to increase our store of scientific knowledge by counting the precise number of hairs on individual dogs at various times on various days, not to test a theory or experiment with a drug to prevent hair loss, but simply to know the canine hair count for its own sake. Even if the information that the biologist collects is true, it has negligible scientific value ... [By contrast] Scientists are interested ... in the form of general the-

ories and laws with *predictive* power. These criteria of scientific excellence—generality and predictive power—and many others besides (such as explanatory power and simplicity) are among the cognitive values of science. They are not the same as truth [emphasis added].[2]

In welcoming, even initiating questioning, science distinguishes itself from dogmatism. Unlike the world of science, the world of dogma forbids its constituency from questioning the beliefs of the system. Followers of dogma are not allowed to study the system, nor examine its veracity or consider whether there are better explanations of the system. They cannot debate the fundamentals upon which the system is based. They are taught unquestioning belief, not to search for truth.

Science, on the other hand, withstands questioning from every quarter. In any forum, all experts' opinions bear consideration, and that consideration will, through consensus, determine the present understanding of truth. German philosopher Jürgen Habermas "stressed the importance of public debate and rational consensus for preventing the domination of society by one group of interests. Consensus suffers inaccuracy when relevant opinions are suppressed. An egregious example was the suppression of Mendelian genetics in Russia in the 1930s."[3] Because of the faulty science of Ivan Vladimirovich Michurin and Trofin Denisovich Lysenko and the politics of Lenin, the Russian government rejected the theory of modern genetics in favor of Lamarckism; the theory that traits can be modified by the environment and then passed on to future generations. Likewise, Nazi Germany rejected Einstein's theory of relativity because he was Jewish.[4] This was dogma not science.

According to Thomas Kuhn, author of *The Structure of Scientific Revolutions* (1962), science distinguishes itself in its ability to solve puzzles, and in order for a field to be called science, its "conclusions must be logically derivable from shared premises."[5] Kuhn observed that scientific communities function by sharing a set of assumptions, techniques, and methodologies, along with a common terminology and worldview, which together become a paradigm.[6] When science turns the magnifying glass on itself, it is called the philosophy of science. The philosophy of science explores such questions as, "What is science?" "What is the value of science?" and "What is the scientific method?" By examining these and other issues, the philosophy of science provides the means for identifying the difference between true science and pseudoscience.

Since earliest times, scientists have recognized that the universe seems to behave with a certain degree of regularity. Science looks for these regularities and calls them *laws* or *theories*. These theories are then offered to explain natural phenomena and to predict future phenomena. Science also relies on axiomatic-deductive reasoning. Scientists set forth principles or *axioms* that are true and self-evident and derive from them more "truths." A theory cannot be accepted into the scientific body of knowledge until it has been tested experimentally and found to be true. In that sense, science is different from other endeavors, such as the humanities. Ideas about the human condition, or expressions of creativity, cannot be scientifically proven. Nonrefutable theories are also not scientific. That is, if something cannot be tested and proven, it is not science.

Science is different from other branches of knowledge in another critical aspect. Scientific progress is contingent upon the expansion of new ideas and the replacement of old ones. Unlike the arts, for example, where the work of a modern-day artist such as Picasso would never replace Michelangelo's *Statue of David*, science advances as scientists replace outmoded theories with improved theories that better explain the world.

The philosopher Imre Lakatos defined science as a paradigm, or model, with a hardcore set of beliefs surrounded by circles of less tenaciously held beliefs. These beliefs are enclosed by a heuristic ring. (Heuristic refers to the educational method in which learning takes place through discoveries that result from investigations made by the researcher.) The core stays the same, while the outer rings change. One experiment cannot prove or disprove a scientific paradigm; rather, new data are always interpreted in light of the core. But the outer rings do change, and these changes fuel scientific progress. In fact, a paradigm is always subject to modification as new knowledge is added; the outer circles are constantly in flux with additions, negations, and modifications.

When modern physics supplanted classic physics in the early 1900s, neither Albert Einstein's general theory of relativity, nor Niels Bohr's explanation of how atoms absorb and radiate energy in terms of quanta did away with Newton's laws of motion and gravitation; they merely modified and reinterpreted this hardcore set of beliefs. Newton's laws were used to put astronauts on the moon in 1969, and continue to be applied to send satellites into orbit. However, Newton's laws cannot explain objects at speeds near that of light (186,000 miles per second). Nor can they explain the effect gravity has on light. The new physics of Einstein and Bohr explained everything Newton's physics explained, plus much more.

Another example of a paradigm shift is Charles Darwin's theory of evolution, which will be discussed in greater detail in the next chapter. Darwin's idea—that all species of plants and animals have evolved from a few common ancestors—revolutionized biological thought when his book, *The Origin of Species,* was published in 1859. Today, geneticists as well as evolutionary and molecular biologists continue to make discoveries about how the process of evolution works at the cellular and sub-cellular level—something Darwin could not explain since the principles of genetics were not yet known. These scientists are adding to and modifying Darwin's theory. But the basic principles introduced by Darwin remain intact.

In the hierarchy of respectability in science, the study of and verification by scientific *theory* is most respected. (Examples from observation and controlled experiments that support or refute the theory are second, and statements by experts are a distant third.) Much of what Einstein and Darwin predicted was based on theory without experimentation. Sir Arthur Eddington stated, "It is also a good rule not to put too much confidence in observational results until they are confirmed by theory."[7] One thing we will do in this book is look at the theory behind using animals to model human disease.

For a theory to become a paradigm, it must be tested repeatedly on an experimental basis. But before concepts, ideas, or technologies reach the test phase, opinions and counter-opinions are debated in a process known as *critical discourse*. For example, Professor Smith thinks a machine can be built that will measure cocaine levels in a person's bloodstream without having to draw a blood sample. Professor Jones says it can't be done. So Professors Smith and Jones discuss the concept. Eventually, one of them will be proved wrong through experimentation. But for the time being, they are content simply to discuss the merits of the concept. Professor Smith may persuade Professor Jones to change his mind, or Jones may succeed in changing Smith's. Or, they may both be wrong. They can only know the answer after they have conducted the appropriate experiments. Kuhn characterized the shift from an opinion based on observation, logic, and rational thought to fact, whether partial or comprehensive, as the move from critical discourse to experimentation.

In the past, before science had the technology to test certain concepts and phenomena, critical discourse provided the final explanation. It remains the final word in arguments when we cannot test theories, such as the existence of God, or when we do not have the data to form an intelligent hypothesis, such as why women suffer from connective tissue diseases more frequently than men. The philosophy of science states "all disagreements about matters of fact are, in principle, open to rational clarification and resolution."[8]

Because science is based on the development of theory, and the proof of theory through experimentation, rigorous standards exist for distinguishing a genuinely scientific theory from one that is pseudoscientific. Kuhn noted that a good scientific theory should fulfill five criteria:

- It must be accurate.
- It must be internally consistent and consistent with other knowledge of the time.

- It should have a broad scope in the sense that it should have implications for phenomenon beyond what it was originally designed to explain.
- It should be simple.
- It should be fruitful; that is, it should yield new knowledge.[9]

A number of demarcation criteria, or parameters, can be used to evaluate a theory in terms of its ability to represent true science. These include:

- *Predictability.* A scientific model allows us to predict subsequent events.
- *Repeatability.* Other scientists can reproduce the phenomena in multiple laboratories and settings.
- *Falsifiability.* There exist conditions that would prove the theory false.
- *Parsimony,* or *Occam's razor.* The theory that explains the phenomena in the simplest way has the most worth. Occam's razor is based on the premise that if two theories explain a phenomenon equally well, then the one with the fewer assumptions or anomalies is true.
- *Mensuration.* Measurements use accepted scales.
- *Heuristic procedure.* The premise that knowledge stimulates more investigation that confirms the knowledge.
- *Generality.* The greater the range of data covered by the theory the better.
- *Consilience.* The data produced conform to known data in other fields.

If a theory or model fails to meet these criteria, it does not qualify as true science. It may be useful for other reasons and it may lead to truth, but it is not science. Many would agree that religion is useful and leads to truth about the spiritual world. Yet it is not science, nor does it claim to be. Pseudoscience, on the other hand, *does* claim to be science. But the fact that it occasionally achieves results that are true does not overcome its inability to meet the demarcation criteria. Philosophers acknowledge that some demarcation criteria may be more successful than others at distinguishing true science from pseudoscience. (For purposes of our discussion, we will focus on the criteria of predictability and falsifiability.)

Science also relies on axiomatic-deductive reasoning. Scientists set forth principles or *axioms* that are true and self-evident and derive from them more "truths." Testing on animals or studying diseases in animals can be done in a scientific manner. However, when the axiom is "animals and humans have so much in common that we can extrapolate the results from animals to humans" then the results are going to be disastrous, not because the scientific method was not followed but rather because the axiom is incorrect. This is why animal-model research is specious science; the axiom is incorrect.

The predictive value of a theory is generally considered to be the most ready and reliable way to separate science from pseudoscience. Only science consistently and accurately predicts a certain outcome that a test will either prove or disprove.[10] When a theory explains phenomena that it did not specifically set out to explain, that is evidence of its validity. Robert Park determines the validity of a theory by asking two questions: "Is it possible to devise an experimental test? Does it make the world more predictable? If the answer to either question is no, it isn't science."

Science also explains after the fact, as does pseudoscience. This is sometimes called *retrospective* explanation, as opposed to *prospective* explanation. Both scientific and pseudoscientific theories are verifiable; both can provide retrospective explanation. According to philosopher Karl Popper, it is easy to find data to support a theory. For example, a believer in astrology can always point to a specific day when her horoscope said something would happen and it did. But only science predicts. Astronomers can predict where Venus will be every day for the next millennium; slightly better results that a horoscope. Rabbits did react to thalidomide as humans but they did not mimic humans in many other cases of drug testing. In other words, in using rabbits to predict birth defects, scientists were unsuccessful.

Popper wrote further that the confirmation of a theory by prediction should count only if the prediction was risky. The prediction, and then proof of the theory by actually doing the experiments to test the prediction, does in part confirm the theory. For example, Einstein's theory of relativity predicted that light would bend during an eclipse. In 1919, Sir Arthur Eddington headed an expedition to study the eclipse of that year. He found that light from stars did indeed bend as Einstein had predicted. Likewise, one should be able to pick any 30-day period of time and find that the astrology-based predictions for all 30 days came true. By selecting *any* 30 sequential days at random as a sample, we are introducing the element of risk. By taking thirty drugs, the side effects, in terms of birth defects, of which are already known and testing how rabbits respond to these drugs, we are also introducing the element of risk.

Popper also said that good theories not only predict happenings, but also *prohibit* things from happening. If, for example, your horoscope said you were going to have a bad day and you won a million dollars in the lottery, obviously your horoscope was wrong. The laws of physics not only predict where the planet Mars will be next year, they also predict that under no circumstances will Mars ever be found between Neptune and Pluto.

In addition to the criteria of predictability, a theory must be evaluated on the basis of falsifiability—the ability to be proven false. Einstein's general theory of relativity was tested and proven by watching light bend during an eclipse. The results could have turned out differently from that which was predicted, and thus the theory could have been proven wrong, or *falsified*. This stands in direct contrast to religious beliefs, such as reincarnation, virgin birth, and resurrection, which are not falsifiable. That is not to say they are not true, only that believing in them is not based in science.

To meet the criteria of falsifiability, a theory cannot be so general as to be encompassed regardless of the results of the experiment. For example, your horoscope, which predicts what kind of day you are going to have, is so vague that many events of the day could be interpreted as having fulfilled the prediction. That is how pseudoscience sometimes escapes falsification—by avoiding testing, or by being tested but making sure the test involves no risk. Pseudoscience also avoids falsifiability by denying that the results of a test falsify the theory. When an alternative medical therapy is tested and proven false, practitioners will often say that one's psyche must be in a particular state in order for the therapy to work. For example, the herb valerian is said to make a "calming" tea that some people use as a mild sedative. Maybe it does for some people. But the anesthetic propofol will put every patient to sleep, every time, regardless of his state of mind.

Pseudoscience also fails to distinguish between causally relevant and casual relationships. The relationship between cigarette smoking and lung cancer is causally relevant; cigarette smoking *causes* lung cancer. When one concludes that a casual relationship is a causal one, it is known as a *post hoc* fallacy. An example of a *post hoc* fallacy is when someone says, "Every time I wash my car, it rains; therefore, my washing the car causes it to rain."

In addition to their ability to meet the standards of predictability and falsifiability, theories are also evaluated on their ability to explain new facts, deal with anomalies, and hold up against alternative theories. A theory is rejected when a better one replaces it or has failed over a long period of time to reconcile the anomalies and is thus considered useless. A good theory is "fruitful" or progressive—in the sense that it leads to new ideas, new connections to existing ideas, and new applications of current theories. Even if a theory explains all the current data, it is considered an intellectual dead end if it is not progressive. Paul Thagard suggests that a proposed theory be deemed pseudoscientific "if and only if, it has been less progressive than alternative theories over a long period of time and faces many unsolved problems." But, "The community of practitioners makes little attempt to develop the theory towards solutions of the problems, shows no concern for attempts to evaluate the theory in relation to others, and is selective in considering confirmations and disconfirmations."[11]

Because a theory proposed by a scientist cannot be accepted as scientific knowledge until it has been verified by other scientists, the "community of practitioners" has a critical role in objectively evaluating the merits of a theory. Therefore, as Thagard points out, one of the ways we can differentiate true science from pseudoscience is by examining whether or not the community of practitioners is holding a theory to the rigorous standards of science. For example, are the practitioners in agreement on the principles of the theory and how to go about solving the problems that the theory faces? Are they trying to explain the anomalies and do they consider the anomalies important? Are they comparing the success of their theory to the success of competing theories? Is the community actively trying to prove or disprove their theory?

All these questions must be explored with diligence and honesty. True science always assumes honest intent—that a scientist will not lie about the results of an experiment just to protect his job, enhance his reputation, or nurture his ego. A scientist once said,

> Anyone who wishes to think rationally should have the habit of thinking coolly, with all affective feelings or sentiments and all emotions parked outside. The heat of the passions, especially if they are strong and violent bodily commotions, cannot help but cause a disturbance or even a distortion of all intellectual work.

Along the same lines, mathematician Mark Kac once said, "a proof is something that convinces a reasonable man, and a rigorous proof convinces an unreasonable man." True, but an unreasonable person may not be convinced regardless of the persuasiveness of the proof. And the easiest way to make a person unreasonable is to make his livelihood dependent on a certain activity. The person whose livelihood is threatened by a new idea is very unlikely to be reasonable. As we have seen, statements made by experts are a distant third (after theory and experimentation) in determining respectability in science. Furthermore, statements against a theory when made by a party with a vested interest in that theory are more reliable than statements made by the same vested interests that are in favor of the theory. In the following chapters, we will see how honest intent—and the notable absence of it—comes up again and again in the debate over the value of animal models in providing information that can be extrapolated to human health problems. As we apply the value of honest intent—and all the other principles of science we have discussed here, including predictability, falsifiability, causally relevant relationships, progressiveness—to the question of animal-model research, it will become clear that animal-model research fails as a scientific paradigm both in theory *and* application.

Science and Animal Models

For most working, tax-paying Americans today, ensuring financial security and maintaining optimum health are top-priority goals. Chances are, you're one of the millions of Americans saving for retirement using proven investment vehicles for obtaining the best return, such as stocks, bonds, real estate, mutual funds, and so forth. Have you ever considered spending half of the money you've set aside for retirement on lottery tickets? Of course not. Any reasonable person knows that throwing half your savings away on a long shot is an irrational and irresponsible investment strategy. Yet, half of all the money invested in healthcare research in the United States today—funds generated through taxes and designed to find new ways to protect your health and that of your family—is spent on the research community's equivalent of the lottery: using animal models to study human disease. (The practice is also known as vivisection.) The odds are about the same, but at least if you have a winning lottery ticket, you'll enjoy a handsome windfall. The payback on animal-model research, in the rare instance when research results are somewhat applicable to humans, isn't nearly as impressive.

So, when it comes to ensuring your family's financial future, would you bet on securities or the lottery? The answer, for any reasonable person, is obvious. By the same token, where would you want to see your healthcare taxes spent? There is the off chance that an animal test will yield applicable results. Alternatively, there are a wealth of nonanimal research modalities available today, such as *in vitro* research, computer simulations, high throughput drug screening, nanotechnology, stem cell research, combinatorial chemistry, epidemiology, human clinical observation, autopsy studies, mathematical modeling, and the human genome project, which have been shown to yield more accurate results, faster and less expensively. The choice should be equally obvious.

Despite its failings, however, animal-model research remains the financial cornerstone of the biomedical research paradigm, largely for reasons that have little to do with the pursuit of scientific knowledge—a factor that has not gone unnoticed by those who have long criticized the animal model. As a result, the debate has raged for years, and promises to escalate as the field of genetic engineering continues to push the envelope of the scientific landscape.

Antivivisectionists oppose animal experimentation for numerous reasons, including those from both an ethical and scientific perspective.

Michael Allen Fox, a professor of philosophy at Queen's University, Kingston, Canada, lists them as follows:

> Today's [anti-vivisection] movement rejects vivisectionist research on these grounds: (a) inapplicability or limited applicability of data to humans owing to cross-species differences; (b) methodological unsoundness (being scientific); (c) dangerous, misleading and harmfulness of results; (d) wastefulness, inefficiency and expense; (e) triviality; (f) redundancy; (g) motivation by mere curiosity; (h) cruelty; (i) availability of alternatives; (j) desensitization of researchers and their coworkers.[12]

In the following chapters, we limit our discussion to those reasons that involve science, addressing the central question from both a *theoretical* and *clinical* perspective: Can knowledge be reliably gained from animals to the benefit of humans? The answer, we will prove, is no. There is much to be lost—and almost nothing to be gained—from using animal models in research, or in harvesting their organs, tissues, and hormones as "spare parts" for humans.[13] This has been true throughout the ages. The *New England Journal of Medicine* stated:

> In the 1400 years between Galen and Vesalius (when autopsies were illegal and experiments on animals thrived), medicine was stagnant, dominated by the belief that illness reflected an imbalance in the four humors—blood, phlegm, yellow bile, and black bile. Life was nasty, brutish, and short, and medical care did not help much.[14]

We submit that:

- Data obtained from animal models in biomedical research, for the purpose of evaluating the safety and effectiveness of pharmaceutical drugs, testing for carcinogens, conducting research on human diseases such as AIDS, and so forth., cannot be reliably extrapolated to humans.
- Even when humans and animals share certain anatomic, physiologic, and metabolic characteristics, such as the presence of mitochondria and the role of serotonin and nitric oxide as transmitters, these characteristics can be studied far more effectively on human cells. That is because, despite gross common characteristics, there will still be microscopic differences between human and animal cells that may mislead researchers.
- Animal organs, tissues, and hormones, such as insulin from cows and pigs, aortic valves from pigs, and monoclonal antibodies produced from animals, are far less effective than the human-derived versions, which are readily available.

Let's first examine the inapplicability of animal data to humans using some historical examples. In animal tests involving thalidomide and penicillin, some animal species were found to react to these drugs in the same way as humans. For example, White New Zealand rabbits gave birth to offspring suffering from the birth defect phocomelia when given thalidomide, albeit only at 25 to 300 times the dosage that was given to humans.[15] But other animal species did not. In 1929, Alexander Fleming tested penicillin on rabbits; because it did not work on a systemic infection in a rabbit, he set the chemical aside, believing it to be ineffective.[16] Mice were just the opposite. Mice reacted to penicillin as humans did, but did not have offspring that suffered from phocomelia after taking thalidomide. And that is the crux of the problem. If every nonhuman animal species never or rarely reacted the same way as humans, or, conversely, if they always or almost always reacted the same way as humans, they might be useful. But since animals react in the same way as humans in an unreliable unpredictable manner, animals are dangerous models for humans. Moreover, it is impossible to know which animal will respond in the same way a human does until it is known how a human responds. It is knowledge *after* the fact—redundant, and therefore useless.

Defenders of animal-model research are quick to point to the few instances when animals did react the same way as humans (mice and penicillin, rabbits and thalidomide), while ignoring the vast majority of experiments that prove the "successes" are the exception to the rule. The contention that animal experiments should continue simply because the results are *sometimes* applicable is scientifically unsound. One must look at the total picture of animal models, not just the occasional incidences when they provide applicable data. As Alexander Pope wrote in "An Essay on Criticism" in 1711, "A little learning is a dangerous thing; Drink deep, or taste not the Pierian spring." Selective learning or taking select data will lead to disaster.

Now for our second point—the study of human cells as opposed to animal cells. Before the dawn of the twentieth century, when so little was known about how the human body functioned, the similarities between humans and animals outweighed the differences. In the nineteenth century, for example, chickens, mice, and humans were all used to prove the germ theory of disease. The results were the same. Both animal dissection and human autopsies were conducted to prove that the heart circulates blood—a discovery that changed the course of medicine. Again, the results were the same. Animals were used to reveal that the body was composed of cells. However, given the tremendous advances in our knowledge over the last 100 years, as well as remarkable developments in technology, the research environment is vastly different from what it was when Louis Pasteur proved the germ theory of disease and William Harvey proved that the heart circulates blood. Using the same tools that were available in the nineteenth century in the face of what is now available can be compared to fighting a high-tech war with weapons from the Revolutionary War. You can debate the relative merits of the various types of muskets used in 1776, but to suggest that any such debate has any relevance to modern-day warfare is ludicrous.

Nowhere is it more critical than in science to use the right tool for the right job. Studying such organisms as yeast has provided us with much knowledge about how cells work at the basic level. Even so, no one has ever attempted to use yeast to study the course of AIDS, because such a different organism would provide no new knowledge of a disease as complex as AIDS. The same principle holds true with animal models. Studying dogs can lead to abundant knowledge about dogs—but not humans.

And finally, with regard to the use of animals as a source for spare parts, it is true that insulin can be derived from cows and pigs, aortic valves can be harvested from pigs, and monoclonal antibodies can be produced from animals. But these products are inferior to the human-derived versions, which are readily available. Human-derived insulin is superior to that which is derived from pigs and cows. It does not cause allergic reactions, as animal insulin often does, which necessitates discontinuing the insulin. Artificial aortic valves last much longer than pig valves; human valves can also be harvested from cadavers. Monoclonal antibodies exhibit a much higher level of purity—and are less expensive—when derived from nonanimal sources.

For all these reasons, animal-model research wastes time, money, and talent—all valuable resources that can be invested in far more effective endeavors for the benefit of human health and safety. Yet the use of the animal model, at best an outdated practice, remains the cornerstone of the biomedical research paradigm. Why?

Why Animal Experimentation Persists

The answers are many and varied. Numerous entities within the scientific and business community have a vested interest in maintaining the status quo of animal model research. That is not to suggest, however, that anyone is "conspiring" to perpetuate animal-model research. A conspiracy is defined as an agreement between two or more individuals to perform together an illegal, treacherous or evil act; in a conspiracy, these individuals agree to commit a crime. Animal-model research is *not* a conspiracy. The vagaries of human nature that keep animal-model research a thriving enterprise are the same ones that have hurt people since the dawn of time: apathy, ignorance, greed, ego, and fear.

One of the most basic reasons animal-model research continues virtually unabated is that people are resistant to change. Newton's first law of motion seems to apply to human as well as planetary bodies. If we have always done something the same way, it is unlikely we will change unless something catastrophic forces us to do so. Ego also plays a significant role. Scientists who experiment on animals and have published the results in hundreds of papers in the scientific literature have their professional reputation—indeed, their entire self-image—on the line. It is self-protection at its most basic level.

Clinicians—doctors who work directly with patients—are simply too busy to take the time to look up where the real medical breakthroughs have originated, and they may well consider it irrelevant. In fact, it does not even occur to them to question the practice of using animal models, even though they do not benefit from them in their practice. Medical students are taught to memorize, rather than to think critically or study the history of their profession. They were taught, by people whose livelihoods depend on using animals, that all the breakthroughs in medicine came from animal studies—a statement they blindly accept as fact.

Researchers, on the other hand, have a slightly different problem. Far removed from patient care, they do not recognize the disconnect between what they are doing in the laboratory and what actually works in clinical practice. They are not unethical, dishonest people. They are simply naive. Werner Hartinger, M.D., a surgeon in the former West Germany, stated in 1989, "There are, in fact, only two categories of doctors and scientists who are not opposed to vivisection: those who don't know enough about it, and those who make money from it."

Perhaps most pervasive, though—and most egregious—is the greed factor. In academia, Ph.D.s are promoted, and thus more highly compensated, based on the number of papers they publish. Animal studies can be concluded much more quickly than human studies. A rat's generation time is weeks, as opposed to decades for a human. So, by the time a clinician publishes one paper, a researcher using animals can publish at least five. The easiest way to publish is to take a concept already published and introduce a new variable, for example, the species of animal used, the drug dosage, or the method of assessing the results. In *The Crisis in Clinical Research: Overcoming Institutional Obstacles*, E. H. Ahrens writes:

> The most research-intensive schools employ only one yardstick for measuring the contributions of the entire staff: the *number* of articles reporting research results. Clearly this is an inappropriate yardstick . . . [17]

One researcher who uses animal models, thinking he was speaking anonymously summed it up: "The chief objective here is to keep us all employed."[18] With quantity seeming to be the primary goal, one can only wonder about the quality—and ultimate value—of these articles. If an article makes a significant contribution to the scientific literature, it will frequently be referred to in later articles, but the journal *Science* reported that approximately eighty percent of all scientific articles appearing in leading journals are never cited more than once in any other publication. Greater than fifty percent are never cited at all.[19]

There are many problems with the scientific journals and consequently with the product they produce. First, because academia encourages quantity not quality and because the same people who are in academia also edit the journals, the journals publish mediocre research. Large difficult subjects tend to be ignored as they take time and will not result in promotion next year whereas irrelevant small studies abound as one of them published per year will result in your obtaining an assistant professorship in five years. Second, because there is such a large volume of mediocre articles no one can keep up with the sheer number of journals in ones specialty much less the articles in the journals; even the important ones. Third, because of the sheer volume the really good articles are spread out over many journals once again making it difficult to keep up with important developments. In summary most articles are simply irrelevant. This can be seen by the number of times other scientists cite them as references.

We can't really fault the average young scientist trying to get ahead. He is just doing what he has been told is the right thing in order for science and medicine to advance. The journal editors and senior staff scientists know better. They have been around long enough to see through the ruse. The reasons they so tenaciously cling to the status quo are complex. Ego and greed play a big role. The process of obtaining medical-research funding in the United States today is a deeply entrenched "old buddy" system, and it is rare that a whistle blower will emerge from this environment. The National Institutes of Health (NIH) is the federal agency charged with the responsibility for allocating taxpayer-generated funds for biomedical research. The most frequent NIH grant—the R01 "investigator initiated" grant—supports both the researcher and the institution where the research is employed. In some cases, the institution receives more money from the grant than the researcher, which it can use at its discretion, whether that means paying the electricity bill or an English teacher's salary. Ahrens continues:

> Since most overhead is brought into the university by a small number of research professors (at Stanford 5 percent of the faculty bring in over one-half of the indirect cost dollars), proposals to reduce research output are not looked on with favor by many university administrators.[20]

It is difficult to know exactly how much of the NIH budget goes to animal-modeled research. Estimates range from thirty to seventy percent. Regardless of the exact amount, however, universities make millions every year. The following is the amount of money (in millions of dollars) for the top ten recipients of NIH grants in 2000. (Note this included both animal-modeled research and nonanimal modeled research)

Johns Hopkins University	$419.4
University of Pennsylvania	$321.3
University of Washington	$302.5
University California, San Francisco	$295.3
Washington University, St. Louis	$279.5
University Michigan	$260.4
Harvard University	$250.4
University California Los Angeles	$243.6
Yale University	$242.8
Columbia University	$226.7[21]

If we take fifty percent as the amount that went to animal modeled research, the top ten recipients alone received over 1.4 billion dollars. (The NIH will not release the exact amount that goes to animal models, so we are forced to estimate.)

The record clearly shows that "The NIH under-funds patient-oriented clinical research."[22] Between 1977 and 1987, only 7.4 percent of the NIH's R01 funding went to basic patient-oriented research. The largest percentage of the awards went to animal experimentation. Ahrens further states, "By far the largest percentage of NIH support for new R01s . . . is awarded to applicants for studies of animal (or microbial) models of human disease. Yet, most experienced investigators realize that animal models of arteriosclerosis, diabetes, hypertension, and cancer are different in important ways from the human condition they are intended to simulate." Only one-third of NIH competing research grant *applications* include human subjects.[23] One U.S. Congressional representative said: "It appears that the [medical establishment] system has changed from one of NIH giving grants for scientific research to one being done solely to get NIH grants."[24]

In 1986, when the president of the Institute of Medicine cautioned that medical research was leaning too heavily on basic animal experiments and not doing enough to support clinical observation, he likened it to the tale of the emperor's new clothes.[25] No one dares call attention to the matter, for fear of direct or indirect retribution. From the standpoint of self-preservation, it's far more prudent to remain silent.

Researchers and universities are not the only ones who profit from animal models. In fact, animal use in biomedical research is a multibillion-dollar business. Animal breeders profit handsomely from the practice. In 1999, mouse sales topped $200 million. That same year, Charles River Laboratories, one of the largest biological supply houses in the industry, reported $140 million in sales of animals; Harlan Sprague Dawley, an Indianapolis-based source of research-related products, sold more than $60 million worth of animals in 1998; Taconic, another major supplier of such products, reported $36 million in sales; and TJL, a not-for-profit, taxpayer-funded corporation, sold $29 million worth of mice. Mice with specific genes missing cost between $100 to $15,000 each.[26]

Suppliers of cages and equipment related to animal-model research have also built themselves a considerable business, as can be seen from these prices for instruments designed specifically for use in animal experiments:

Stereotaxic device for rats	$4,500.00
Stereotaxic device for cats and monkeys	$7,215.00
Stereotaxic device for dogs	$8,580.00
Metabolic gas monitor	$27,300.00
Flat treadmill for rodents	$9,600.00
Incapacitance analgesia meter	$7,300.00
Sliding microtome	$9,975.00
Muromachi microwave fixation system for humane killing with immediate de-activation of brain enzymes	$70,200.00

Even the media profits from animal experimentation. "Editors want the medical miracle."[27] Television and newspaper reports on "breakthrough" drugs may exaggerate their effectiveness and minimize the side effects, but no matter—the sensational reports sell papers and increase ratings. Not a week goes by when we are not regaled with stories suggesting that "a cure" for Alzheimer's, cancer, or the like has been "discovered." A perusal of the details reveals that the "cure" was wrought in genetically modified mice and that other such "cures" have not translated to humans.

Pharmaceutical companies also have a hand—and an exceedingly large one at that—in perpetuating animal use in research. Their biggest liability is when a drug damages a patient, and they know that all drugs will injure someone, sometime, somewhere. If a pharmaceutical company can say to the jury that they did what the law required—"proved" a drug's safety in animals—then they feel that they've gone a long way toward getting themselves off the hook for a multibillion-dollar award. Hugh LaFollette and Niall Shanks explain this in "Animal Experimentation: the Legacy of Claude Bernard," published in the *International Studies in the Philosophy of Science*.[28]

As one prominent animal researcher explains it: "Judges and juries may not be able to evaluate the scientific implications of primate studies, but they are favorably impressed when a manufacturer appears to have done more than the required minimum by testing his product on pregnant primates." But the same researcher also notes that: "Animal tests, however extensively or carefully done, can never establish human safety regarding teratogenic risks from some new chemical or physical agents. The numbers of potential human exposures will usually far exceed the numbers of animals used in such tests, quite apart from the impossibility of precise extrapolation of data from one species to another."[29]

For corporate giants, animal studies are a legal safety net, and they have no incentive to change the status quo—even though eliminating animal testing would save them millions of dollars, maintaining it saves them even more. The current system works in their favor, providing a reassuring level of liability protection. Why rock the boat?

Guilt keeps the animal experimentation machine operating as well. Many people love animals—including some who use them in research. These researchers truly believe that they are doing the right thing; if they remove their psychological blinders to see the animal model for what it really is, the ensuing feelings of guilt would be overwhelming. More than once after one of our university lectures, we have had a researcher ask us with tears in her eyes, "Drs. Greek, if what you say is true, why have I killed all those animals?" No one wants to feel that kind of emotional pain.

Obviously, the above is a simplified version of reality. Many other issues factor in. Such as, if the way biomedical research was conducted were to change dramatically the very social order in universities and other research institutions would also change dramatically. The consequences of this social restructuring would be impressive. While we do not have space enough to explore all the other reasons animal models persist, suffice it to say money is still a big factor.

Science versus Pseudoscience

While animal breeders, manufacturers of cages and equipment, and pharmaceutical companies all share part of the blame for fueling the animal model industry, the scientific community is "ground zero" for animals used in research. That is why when people learn the truth about animal-modeled research, they are quick to blame science, when in fact science, in providing us the best way to understand the material world, is not the problem at all.

As doctors and scientists, we will always defend science, rational thought, and the trial and error process of the scientific method. History has taught us that rational thought is preferable to irrationality, that the scientific method is an excellent way to develop reason, and that intellect must lead in the pursuit of truth. To discount the value and importance of science because an unsound, outdated practice proliferates under the guise of scientific investigation is a huge mistake. It's like throwing the baby out with the bath water. There is good science, and there is bad science—or what we call *pseudoscience*. Animal-model research is an example of pseudoscience or what we call specious science; an argument that sounds true but is in fact false. Using animals to model humans sounds reasonable until one probes further into the practice. But it is not the only example, as we shall see in a moment. Pseudoscience is increasingly pervasive in our society today, largely because of the public's misguided perception of science.

Science has been under attack recently because it appears to have failed us on a number of key issues, including the apparent failure of scientists to predict mad cow disease, the apparent lack of consensus over global warming, or the recent multiple failures of NASA to put a probe on Mars. The world looks askew at science because of Bhopal, Chernobyl, the *Challenger* disaster, it's failure to cure AIDS, controversies surrounding stem cells and cloning, and the atomic bomb. Such failings and lack of consensus do not inspire public confidence. The public's trust in science has also been shaken by their belief that in any lawsuit involving science, there will always be a scientist who is willing to say whatever he is paid to say. Almost everyone has heard of a court case where an expert's opinion was sold to the highest bidder, the silicon breast implant, the tobacco industry's claim that smoking was safe, and Bendectin birth defect lawsuits being some of the most widely publicized examples. When people see that almost anyone can be considered a medical expert, they lose respect for the medical profession—and for science in general.

Such deep cynicism is a relatively recent phenomenon. There was a time when people had more faith in the objectivity of the application of science, and that the truth of Sir Isaac Newton's laws of motion would always stand, whether or not such laws were politically correct. But that was before scientists assured citizens in the United Kingdom that their beef supply was safe (it wasn't), that cholesterol had nothing to do with heart disease (it does), that smoking was not addictive (it is), that vitamin "O" existed (it doesn't), and that laetrile could cure cancer (it can't). Revelations that scientists who made false or misleading public statements were often paid to do so by vested-interest groups, such as tobacco companies or the livestock industry, have served only to fan the flames of skepticism.

One of the most revealing examples of why people are skeptical of science is the very public and vigorous disagreement between scientists over the greenhouse effect. (The greenhouse effect is the warming of the lower atmosphere and surface of the earth as a result of a complex process involving sunlight, gases, and particles in the atmosphere.) Robert Park, writing in *Voodoo Science: The Road from Foolishness to Fraud*, addresses the issue:

> If scientists claim to believe in the scientific method, and if they all have access to the same data, how can there be such deep disagreements [about things like global warming] among them. . . . What separates them are profoundly different political and religious worldviews. In short, they want different things for the world. The great global warming debate, then, is more an argument about values than it is about science. . . . Most scientists, however, were exposed to political and religious worldviews long before they were exposed in a serious way to science. They may later adopt a firm scientific worldview, but earlier worldviews 'learned at their mother's knee' tend to occupy any gaps in scientific understanding, and there are gaps aplenty in the climate dispute. This sort of dispute is seized upon by postmodern critics of science as proof that science is merely a reflection of cultural bias, not a means of reaching objective truth. They portray scientific consensus as scientists voting on the truth. That scientists are influenced by their beliefs is undeniable, but to the frustration of its postmodern critics, science is enormously successful. Science works.[30]

Unfortunately, the fact that science does indeed work is often lost on the public, perhaps because what is presented in the media and elsewhere is largely not science. There is a world of difference between what is published in scientific journals and what the media reports. For example, the discovery of cold fusion (and the scientists who "discovered" it) was widely trumpeted in newspapers, television, and radio before it was debunked, and even after the scientific community revealed cold fusion to be a fraud, the media clung tenaciously to it for weeks afterward. Cancer "cures" found in animals are in the news almost daily, but people still die from cancer. Why? Because news like this—truthful or not—sells.

Some of the problem stems from people failing to distinguish between what scientists state *conclusively,* and what they state *tentatively.* Scientists will say conclusively state that no condition can violate the laws of thermodynamics. And they will say conclusively that there is no perpetual motion machine. However, scientists can, and do, argue about the safety of genetically modified organisms, because, at least at the time of this writing, there is not enough data to say conclusively one way or the other. As Park continues in *Voodoo Science*:

> Only a tiny fraction of all scientific research is ever covered by the popular media, however, and most scientists go through their entire

career without once encountering a reporter. New results and ideas are argued in the halls of research institutions, presented at scientific meetings, published in scholarly journals, all out of public view. Voodoo science, by contrast, is usually pitched directly to the media, circumventing the normal process of scientific review and debate. We saw this in the case of the Newton Energy Machine, the Patterson cell, and the cold fusion claims of Pons and Fleishmann. The result is that a disproportionate share of the science seen by the public is flawed. The reluctance of scientists to publicly control voodoo science is vexing. While forever bemoaning general scientific illiteracy, scientists suddenly turn shy when given an opportunity to help educate the public by exposing some preposterous claim.[31]

Regrettably, the public's "general scientific illiteracy" has lent credence—however unjustifiable—to the claims of postmodernists, who argue that so-called inconsistencies in science somehow make it an incorrect way of gathering knowledge about the material world. When the door is left open to question the entire process of science, trouble is certain to follow. Dr. William F. Williams of the University of Leeds has stated:

> There is an increase in the public's attraction to the whole variety of phenomena and activities that lie outside the normal concerns of orthodox science. Astrology columns and horoscopes in the press have proliferated and can now be found in some quality newspapers, which, only a year or two ago, would have strenuously denied them admission. . . . Political leaders turn to astrologers and spiritualist mediums for advice about matters of national, and even international, importance. Cults, always present in society, appear to be increasing both in number and in strangeness and are treated with more respect than was once the case. In these and in other ways, the turn from the second to the third millennium is seen as flagging a change from the rational (appealing to reasoned argument) to the irrational or, perhaps, as precipitating some major event for which this preoccupation with what is seen to be the spiritual will provide some sort of insurance. This is puzzling behavior. For many years, there has been a growing reliance throughout the world on science and its close relative, technology, both representing an essentially rational outlook. Now we appear to be regressing.

The public is easily dissuaded from scientific arguments in part because our educational system does not emphasize critical thinking; hence, people do not understand the importance of science and are not prepared to think through scientific arguments. Culturally we are geared toward escapism and entertainment. Scientific concepts can be quite difficult to comprehend, and they often demand diligent study. Moreover, the trend toward specialization in science fields keeps decision making remote. Norman Levitt, professor of mathematics at Rutgers University writes in *The Flight from Science and Reason*:

> Only a woefully small number of intellectuals have actually engaged in the rigid kind of mental calisthenics where principles of logical inference are rigidly applied, ambiguities of language sifted out, unstated assumptions kept from the premises, words prohibited from sliding unannounced from one meaning to another, and appeals to emotion or cultural prejudice or moral indignation despised. This is an exercise well worth going through, just to see that it can in fact, be done. . . . The real value, as I see it, of even a modest mathematical education is that it breeds a certain salutary impatience, a distaste for intellectual flatulence, for other pseudotheorizing, for argument by browbeating. It breeds a certain shrewdness, as well, in all sorts of odd corners of modern life. It helps purge the staleness, the laziness, the careless propensity to accept unexamined clichés, from one's thinking, simply because it provides a rich array of mental patterns and the habit of looking for instances where they are applicable.[32]

In *Logic for the Millions*, Alfred Mander states:

> Thinking is a skilled work. It is not true that we are naturally endowed with the ability to think clearly and logically—without learning how, or without practicing. People with untrained minds should no more expect to think clearly and logically than people who have never learned and never practiced can expect to find themselves good carpenters, golfers, bridge players, or pianists.

Left without adequate critical thinking skills, people become vulnerable to that which makes them feel comforted and in control of their lives. The stage is then set for flights of fancy into the world of pseudoscience, where the germ theory of disease, Boyle's Law, genetics, and Newton's laws are placed on the same level as biorhythms, pyramids, faith healing, and Bigfoot.

A 1990 Gallup poll revealed that fifty-two percent of Americans believe in astrology, seventeen percent in witches, and forty-two percent in communications with the dead. In addition, sixty-seven percent said that they had had a psychic experience. Living as we do at the apex of humanity's scientific accomplishments, it is ironic that people appear to be returning in droves to the Dark Ages of alchemy and irrationality, where medicine was more magic than science. Philosopher Paul Thagard states:

> . . . society faces the twin problems of lack of public concern with the important advancement of science, and the lack of public con-

cern with the important ethical issues now arising in science and technology, for example around the topic of genetic engineering. One reason for the dual lack of concern is the *wide popularity of pseudoscience* and the occult among the general public. *Elucidation of how science differs from pseudoscience is the philosophical side of an attempt to overcome public neglect of genuine science* [emphasis added].[33]

Alternative Medicine and Animal Models

The dramatic growth of so-called alternative medicine in recent years is a good example of this phenomenon. It provides an enlightening case study of how pseudoscience can gain a level of credibility and respectability far beyond its ability to measure up to the standards of true science. As such, it is worthwhile taking a moment to examine alternative medicine and see how it—and the dangers it presents—offers interesting parallels to another pseudoscience: using animal models in biomedical research.

First, some definitions. Alternative therapy—treatments that are unproven because they have not been tested or were tested and found ineffective. Complementary therapy—supportive therapy used to complement traditional therapy. Complementary therapy does not cure but promote a sense of well-being and reduce stress in some people. Traditional treatment—treatment that has been tested by scientific methods and found efficacious.

There is no such thing as alternative medicine. There is only scientifically proven evidence-based data and therapy and treatments that have not been scientifically proven. Traditional medicine does not claim to have all the answers; it does claim to know how to ask the questions. Alternative medicine is a pseudoscience. Science is a field of knowledge that is based on evidence obtained through systematic methods of study in which scientists make observations and collect facts. Knowledge can be defined as truly scientific only through being repeatedly tested experimentally and found to be true. For example, penicillin cured infections 60 years ago, 50 years ago, and so on including the present. It has been tested millions of times—with the same results.

Traditional medicine is true science because it involves diagnosing and treating patients through the application of methods and therapies that have been tested and found to be efficacious. Alternative medicine, which includes a wide range of healing practices such as acupuncture, chiropractic, naturopathy, herbalism, homeopathy, and even faith healing and psychic healing, cannot be considered science because they rely heavily on *anecdote, testimonials,* and *belief* rather than *evidence* obtained through careful study and repeated experimentation. The quality of evidence in medical science is determined in accordance with the following system, which ranks the evidence from an experiment (or series of experiments) from best to least as follows:

- Multiple randomized, prospective, double-blind, multi-institutional studies: These studies involve a large number of patients in several hospitals who are undergoing a treatment. Researchers randomly divide experimental subjects into a control group, which is given a placebo (a substance containing no medication), or a drug whose effects are known, and an experimental group, which receives the treatment under study. Neither the physician administering the substances, nor the patients in the study, know which group they are in.
- One randomized, prospective, double blind, multi-institutional study.
- One randomized study.
- One well designed, controlled trial without randomization.
- One well designed cohort or case-control analytic study.
- Evidence from multiple time series.
- Dramatic results from an uncontrolled study (for example, the original human testing of penicillin).
- Opinions of respected authorities based on clinical experience, descriptive studies, or reports of committees.[34]

Nowhere on this hierarchy does one find anecdotes, testimonials, and beliefs—all upon which alternative medicine relies upon almost exclusively to "prove" its validity. Anecdotes and testimonials are alternative medicine's stock in trade. Unproven treatments are often promoted as successful because someone knows someone who claims his ailment was improved by using it. Many times, alternative medicine is credited for curing a disease that was actually cured by conventional medicine, such as when a cancer patient uses aromatherapy while undergoing chemotherapy for breast cancer, and then credits the aromatherapy for her remission. (Just as when the animal model vested-interest groups claim that because a researcher experimented on animals, the animal models caused the breakthrough, when in fact the breakthrough had nothing to do with the animal models.) Claims that an alternative medicine therapy cured an ailment that would have eventually gone away without any treatment are common as well. It is very easy for proponents of alternative medicine to take advantage of the fact that many illnesses and conditions, such as pain, have a significant subjective component and are influenced by many factors. Placebos do alleviate pain for some people some of the time. Nevertheless, claims such as "it worked for me!"—and therefore will work for others—have no basis in science.

If acupuncture is tested as penicillin, appendectomies, antihypertensive medications and MRI scans have been, and proves effective, then and only then can it be said that acupuncture has been scientifically proven. (We will say the same about animal models.) No doubt, some practices that are considered alternative medicine today will, tomorrow, be shown effective in scientific studies. Then they will no longer be alternative but will fall under the umbrella of traditional medicine. The reason we have such a high standard of medical care today is due to the fact that scientists have taken untested methods and tested them repeatedly, discarding those therapies that cannot be proven efficacious. Conventional medicine does not dismiss unproven methods out of hand; treatments that are curing disease and prolonging lives today all started out as unproven therapies. The difference between science and pseudoscience is the use of the scientific method—vigorous, repeated testing—in determining what works and what doesn't and why.

Although we have made great progress in understanding the functions of the human body and the nature of disease, we have not yet discovered the answers to all our questions. True, a cure for most cancers continues to elude scientists, and such debilitating illnesses, as Lou Gehrig's disease, muscular dystrophy, multiple sclerosis, Alzheimer's disease, and Parkinson's disease remain largely a mystery. Still, it is science via methods such as the randomized, prospective, double-blind, multi-institutional study, the results of which are thoroughly analyzed using sophisticated statistical methods and basic science research using human-derived tissues, that offers the best chance we have to solve these medical mysteries.

No theory can be accepted as part of our body of scientific knowledge unless it has been verified by the studies of other researchers. Therefore, none of the unproven medical therapies collectively known as alternative medicine will be accepted by conventional science until they are tested in multiple double-blind studies and proven to be effective. All the anecdotes in the world cannot substitute for the scientific method because anecdotes are based on subjective experiences, and therefore cannot qualify as objective data. Science does not claim to have all the answers, but it rightly claims to know how to ask the questions that separate fact from fiction.

Even though there may be more fiction than fact embodied in alternative medicine, what harm can there be in, say, using aromatherapy to counter some of the uncomfortable side effects of chemotherapy? Nothing, really, in and of itself. But that is not to say that certain alternative medical therapies are not without risk—some of them quite serious. Those who believed, erroneously, that "natural" compounds are always safe should consider the recent example of the Chinese herb *Aristolochia fangchi*. Used as a "natural" weight loss aid, it has already been found to cause kidney damage and is now also thought to cause cancer.[35]

Recently, hydrazine sulfate, an unregulated remedy for cancer marketed on the Internet, caused the death of a patient.[36] Dietary supplements containing ephedra alkaloids (also known as *ma huang*) have been shown to cause strokes, heart attacks, and high blood pressure.[37] We know of no evolutionary reason to believe that naturally occurring chemicals act on the human body to produce maximum benefit with little or no side effects. Arsenic is a naturally occurring chemical.

Animal models and alternative medicine have much in common. The potential for serious physical harm—even death—is a direct, obvious risk but, there is an equally devastating, though less obvious indirect risk. Patients who take the path of alternative medicine—who, for example, take shark cartilage rather than standard chemotherapy to treat their cancer—make the choice to ignore tested and proven therapies that could help them. All too often, when the alternative therapy doesn't work, and the patient seeks out conventional therapy, the disease has progressed to the point where it is far more difficult to treat, or even too late to treat at all. Or, because they have invested all their financial resources on unproven therapies, they have no money left to obtain conventional medical treatment. The same applies to the animal model. The United States does not have unlimited resources to spend on biomedical research. Every dollar that goes to animal models is a dollar that does not go to *in vitro* research, autopsies, technology research, epidemiology, chemistry or math research. So if there is an *in vitro* research project that may yield clues about Alzheimer's, it will not be funded so that an animal model of Alzheimer's can be funded. Further, the data that comes from the animal model may result in treatments being applied to humans that kill or maim them, as we will see.

The alternative medicine industry also benefits from the gullibility of the general public, including legislators, who are not educated in the basic tenets of science and therefore susceptible to unsubstantiated claims of cure-alls. In 1992, Senator Tom Harkin of Iowa introduced legislation creating the Office of Alternative Medicine at the National Institutes of Health. Harkin claimed to have been cured of allergies by swallowing large amounts of bee pollen. Later, the person who sold Harkin the bee pollen was forced by the FDA to pay a $200,000 fine for false advertising.[38] Without a basic understanding of science, legislators are apt to rely too much on their own personal experience or the opinions of others, and vested-interest groups are only too eager to sell their side of the story. Besides, money talks. It was the natural dietary supplement industry's massive lobbying effort on Capital Hill that was largely responsible for Congress passing the highly favorable (to the industry) Dietary Supplement and Health Education Act in 1994, which exempts all natural dietary supplements from having to be tested for purity, effectiveness, or safety in accordance with FDA (Food and Drug Administration) regulations. The passage of this act, which is clearly not in the

interests of the American public, demonstrates how politics and ignorance get in the way of public health. The great latitude with which
practitioners of alternative medicine are given to operate, as well the
reasons for the growing popularity of alternative medical therapies in
today's society, bear a striking resemblance to the problem of using
animal models in biomedical research. Neither alternative medicine nor
animal-based research is grounded in sound scientific principles. Both
these pseudosciences thrive in a culture that is mistrustful of science, yet
fascinated by seances. Proponents of alternative medicine, like those who
support animal-based research, know how to use the public's lack of
understanding of the key tenets of science to serve their own interests.
The potential damage to society is the same: public health and safety is
compromised, and precious resources are used up pursuing meaningless
theories and wasteful projects.

All of us want a life in which we make the right choices. We expect
the same from our leaders. Since each and every one of our decisions has
a consequence, how can we increase our odds of making the right
choices while decreasing the likelihood of making the wrong ones? Intuition is frequently used as a guide in making choices. But many things
proven to be true by science are counterintuitive. Intuitively, one would
not suspect that an infinitely long fence could surround a finite area of
field but the condition can exist. Science is successful because it discovers
and describes realities about the material world that are completely
counterintuitive, yet true. Gross and Levitt state in *Higher Superstition,*

> The dissecting blade of scientific skepticism, with its insistence that
> theories are worthy of respect only to the extent that their assertions
> pass the twin tests of internal logical consistency and empirical
> verification, has been an invaluable weapon against authoritarian-
> ism of all sorts, not least those that sustain social systems based on
> exploitation, domination, and absolutism.[39]

True science dismisses absolute obedience to authority. As a result, it
has been successful in developing and testing theories that have advanced
human understanding of the material world around us, while disproving
theories and principles based on untenable beliefs. This is the value—
and the uniqueness—of science. As scientists continue to accumulate
more information, theories are constantly being revised and updated,
and scientific knowledge is always growing and improving as outdated
theories are replaced with new ideas. One of the outmoded theories
that need to be replaced is the idea that knowledge gained from animal
models is applicable to humans. In the following pages, you will see
how animal-model research fails as a scientific paradigm from a
theoretical viewpoint. The remaining chapters provide numerous case
studies and examples of how the failure of animal-model research has

actually harmed humans and delayed medical progress in internal medicine, the development of medications, surgery, pediatrics, and diseases of the brain.

In its conceptual weaknesses, inconsistencies in logic, and inability to provide relevant data, animal-model research is as much a pseudoscience as alternative medicine, astrology, and other similar endeavors. In the remaining chapters you will be able to compare animal-modeled research to the points made in our above discussion of alternative medicine. To fully appreciate exactly how animal-model research fails as true science requires an understanding of exactly what science is—and what it is not—and how that explanation of science applies to the animal model. In the next chapter, we explore the animal model and compare it to our previous definition and explanation of science.

chapter 2

The Theoretical Basis for the Failure of the Animal Model as a Scientific Paradigm

To stumble twice against the same stone is a proverbial disgrace.

—Marcus Tullius Cicero (106–43 B.C.E.), Roman Orator and Statesman

Animals are similar to humans, almost the same. The chemical in antifreeze, ethylene glycol, is almost the same as ethanol, the favorite beverage of college students. But ethylene glycol kills. Two chemicals, almost identical yet one can be fun and the other deadly. The chemist Primo Levi in his autobiography, *The Periodic Table,* warns against using the "almost the same" in chemistry:

I thought of another moral . . . that one must mistrust the almost-the-same . . . the practically identical, the approximate, the or-even, all surrogates, and all patchwork. The differences can be small, but they can lead to radically different consequences, like a railroad's switch points; the chemist's trade consists in good part in being aware of these differences, knowing them close up, and foreseeing their effects. And not only the chemist's trade.[1]

The same warning should be heeded by those using animal models.

Proponents of animal-model research are quick to credit the paradigm for almost exclusively bringing about the advanced state of medical knowledge we enjoy today. Their claims are extraordinary—and extravagant:

". . . we cannot think of an area of medical research that does not owe many of its most important advanced to animal experiments."[2]

"Every major medical advance of this century has depended on animal research."[3]

"Virtually all medical knowledge and treatment—certainly almost every medical breakthrough of the last century—has involved research with animals. There is a compelling reason for using animals in research. The reason is that we have no other choice. . . . There are no alternatives to animal research."[4]

"Virtually every major medical advance of the last 100 years (as well as advances in veterinary medicine) has depended on research with animals. Animal studies have provided the scientific knowledge that allows health care providers to improve the quality of life for humans and animals by preventing and treating diseases and disorders, and by easing pain and suffering. Knowledge gained from animal research has contributed immeasurably to a dramatically increased human life span."[5]

". . . virtually every advance in medical science in the twentieth century, from antibiotics and vaccines to antidepressant drugs and organ transplantation, has been achieved either directly or indirectly through the use of animals in laboratory experiments."[6]

". . . research with animals has made possible most of the advances in medicine that we today take for granted . . ."[7]

"There is no question that most medical progress—perhaps all, in fact—has been attained through knowledge derived initially from experiments in various animal species."[8]

Extraordinary claims, to be sure. But are they true? It is our position that animal-model research is scientifically untenable both in its theory and application, and that claims, such as those above are unsupportable. As such, the study of animal models, though represented as science, is in fact pseudoscience because it violates the criteria that form the foundation of true science. In this chapter, we will discuss how the animal-model research paradigm fails from a theoretical viewpoint. Here, we will apply some of the principles discussed in the previous chapter to demonstrate the profound lack of a sound theoretical basis for the use of animal models in the study of human disease.

The Animal Model and Evolution

First, let's establish exactly what we mean by "animal model." There are basically eight ways animals can be used in medicine, biology, and biomedical research:

(1) Animals as spare parts (for example, heart valves).
(2) Animals as factories (for example, insulin and monoclonal antibodies).
(3) Animals as models for human disease.
(4) Animals as test subjects (for example, drug testing, carcinogen testing).
(5) Animal tissue to study basic physiological principles.
(6) Animals for dissection in education.
(7) Animals as a modality for ideas (for the purpose of heuristic procedure).
(8) Using an animal to study a disease or condition for the benefit of the same species but not the benefit of the individual animal being studied. (We will leave aside this example, as it is an ethical, not scientific issue.)

In the laboratory, animals are used as exemplified in 1–7. Michael S. Rand, DVM, Chief, Biotechnology Support Service at the University of Arizona-Tucson stated in 1999:

> The aim of using animal models in biomedical research is to reconcile biologic phenomena between species, i.e., we wish to examine systems existing in one species and extrapolate knowledge to another.[9] The term *model* in this usage denotes not "a small version of the thing itself" nor "a blueprint or design of the thing itself." A *model* here is a device that enables us to conceptualize unfamiliar phenomena by *analogy* to qualitatively different but familiar phenomena.[10]

Early animal experimenters assumed that if one type of tissue in two different species performs the same *function*—say, respiration—then the *causal* mechanism of the function is the same. This concept has led researchers to maintain that animals are accepted *causal analogical models* (CAMs) and can be used to study human disease. The reasoning process for CAMs is called *causal analogical reasoning* (CAR). Causal analogies are a subset of analogy arguments in which causal assumptions arise based on the model. In their book *Brute Science: Dilemmas of Animal Experimentation,* (which we recommend highly) Hugh LaFollette and Niall Shanks explain that the first necessary condition for a thing to be considered a CAM is this: "X (the model) is similar to Y (the object being modeled) in respect {a . . . e}. X has additional property *f*. While *f* has not been observed directly in Y, likely Y also has property *f*."[11] So, if drug Z *causes* death in an animal model (for example, penicillin kills a guinea pig), animal experimenters reason *by analogy* that it will also cause death in humans.

LaFollette and Shanks state that:

CAMs must satisfy two further conditions: (1) the common properties [$a, \ldots, e$] must be causal properties which (2) are causally connected with the property {f} we wish to project—specifically, {f} should stand as the cause(s) and effect(s) of the features {$a, \ldots, e$} in the model.[12]

In other words, CAMs must have: (1) common causal features, (2) causal connections between the features, and (3) no causally relevant disanalogies. However, the same function can be arrived at by different evolutionary pathways and different causal mechanisms. Birds ventilate differently than humans, so the causal mechanism is different. But the ventilation accomplishes the same function—respiration—in both species. This is called causal/functional asymmetry, and the theory states: "Although we cannot infer similarity of causal properties from similarity of functional properties, we can infer differences in causal properties from differences in functional properties."[13]

The causal/functional asymmetry theory implies that causal mechanisms may differ between species. Therefore, causal disanalogies mandate caution in extrapolating data between species, which poses a huge theoretical problem for those who defend the animal-model paradigm. Yet the theory of causal/functional asymmetry is firmly rooted in the single organizing principle of modern biology: the theory of evolution. In fact, evolutionary biology supports and explains the biochemical reasons for questioning the extrapolation of the results of experiments on animals to humans. Evolutionary theory shows us why it *appears* we can use animal models as well as the reason why in reality we *cannot*. It clearly demonstrates the shortcomings of the concept that animals can be used to model for humans. As such, the animal-model theory fails to meet the first criteria of respectability in science: the study of and verification by scientific theory, in this case evolution.

In his book *Evolutionary Biology,* D. J. Futuyma stated, "Evolution . . . is the central unifying concept of Biology. By extension, it affects almost all other fields of knowledge and must be considered one of the most influential concepts in Western thought."[14] Stephen Jay Gould wrote, "[Evolution] is as well documented as any phenomenon in science, as strongly as the earth's revolution around the sun." While evolution is referred to as "theory," it is supported by such a plethora of evidence from numerous scientific fields that it has become accepted as fact. The notion that God created each species 10,000 years ago in six days, in essentially the same form they exist today, is no longer considered plausible.

Evolutionary theory holds that all living things on earth evolved from a single form of life that inhabited the earth millions of years ago. Over eons of time, this basic life form evolved into multiple phyla, classes and species through a branching process known as *speciation*. Because of its common heritage, all life on earth has common characteristics. But millions of years of adaptation and natural selection have created ten million plant and animal species as diverse as wart hogs, Venus flytraps, chimpanzees, humans, *E. coli*, and poison ivy. Evolution allows us to categorize life forms into groups known as *species*. A species like Homo sapiens (human) will have characteristics that are unique to it. It will also have characteristics that it shares with other species, such as *Drosophila melanogaster* (the fruit fly) or *Pan troglodytes* (the chimpanzee).

In 1859, the renowned naturalist Charles Darwin introduced his theory of evolution in *The Origin of Species*. At the time, Darwin drew his ideas from a number of sources, including his personal observations as a member of a scientific expedition on the *H.M.S. Beagle* from 1831 to 1836, the geological theory of the British scientist Sir Charles Lyell, and the population theory of the British economist Thomas Robert Malthus. However, Darwin, given the time period he lived in, had no way to explain how or why natural selection—where the individuals who were most suited to their environment tended to survive and pass on their characteristics to their environment—took place.

Developments in the field of genetics and molecular biology eventually filled in the gap left in Darwin's theory, with the discovery of the hereditary raw material of all organisms: deoxyribonucleic acid, or DNA. Genes, the basic units of inheritance, which determine an organism's characteristics, are composed of DNA. Genes are located in chromosomes, which are located in the nucleus of a cell. DNA is a nucleic acid sequence composed of *phosphate*, a sugar called deoxyribose, and compounds called *bases*. There are four different bases—adenine (A), guanine (G), thymine (T), and cytosine (C). A always pairs with T, and G always pairs with C. (Hence the title of the movie *GATTACA*.) Together, these components are arranged in chemically bonded units called *nucleotides*; the nucleotides bond to each other to form long chains called *polynucleotides*. A DNA molecule consists of two chains of polynucleotides arranged in a double helix—the familiar, spiral staircase everyone has seen. The sides of the ladder are the polynucleotide chains of phosphates and sugars. The base pairs of AT and GC are the rungs.

When a gene is "expressed," or switched on, that portion of the double helix unwinds, and the base pairs separate. That is, the G separates from the C, and the A from the T. The single strands are then "read" and copied to make ribonucleic acid, or RNA. RNA is composed of nucleotide bases like DNA. The RNA is then read and "translated" into amino acids, with each triplet of nucleotides (or codon) coding for one amino acid. Amino acids then combine to make proteins, which are essential to all plant and animal life because it is protein activity that dictates the form of all living things.

DNA sequences either side of the gene determine where the "reading" should start and stop; so the genetic material is actually composed of "structural" genes coding for proteins and "regulatory" genes that determine if the structural gene is turned on or off. These are the switches.

Evolution occurs at the molecular level by the substitution of one nucleotide for another. A change in a single nucleotide can reorder the sequence of amino acids and hence make a different protein. The four DNA base pairs, A, T, G, and C, code for twenty universal amino acids, and the myriad different combinations of these form the multitude of proteins that create the extraordinary variety of life we see on earth. Thus all species, from insects to humans, plants and animals, follow the same design; not only are they formed from the same DNA units (A, T, C, and G), they are also assembled using the same process. But while all plant and animal species share the same genetic material, it is the *composition*, or arrangement, of this genetic material that makes all the difference.

Lewis Wolpert, in *The Triumph of the Embryo*, explains:

> Compare one's body to that of a chimpanzee—there are many similarities. Look, for example, at its arms or legs, which have rather different proportion to our own, but are basically the same. If we look at the internal organs, there is not much to distinguish a chimpanzee's heart or liver from our own. Even if we examined the cells in these organs, we will again find that they are very similar to ours. Yet we are different, *very* different from chimpanzees. Perhaps you may wish to argue, the differences lie within the brain. Perhaps there are special brain cells which we possess that chimpanzees do not. This is not so. We possess no cell types that the chimpanzee does not, nor does the chimpanzee have any cells that we do not have. The difference between us and the chimpanzees lies in the spatial organization of the cells.[15]

One reason for the difference between species vis-à-vis the spatial organization of the cells lies within the genes. Wolpert continues:

> The face develops from a series of bulges in the head region and at early embryonic stages it is not easy to distinguish dog from cat, mouse from man. The differences in facial features are very dependent on just how much these bulges grow. One can begin to imagine how genes could control such changes in growth rates at different positional values. The key changes in the evolution of form are in those genes that control the developmental programme for the spatial disposition of cells. The difference between chimpanzees and humans lies much less in the changes in the particular cell types— muscle, cartilage, skin, and so on—than in their spatial organization. Direct confirmation of this comes from studies which compare the proteins of humans and apes. If we look at the genes that code

for the average "housekeeping" proteins—proteins that function as enzymes or provide basis cell structure and movement—the similarity between chimpanzees and humans is greater than ninety-nine percent. The difference must reside not in the building blocks but in how they are arranged, and these are controlled by regulatory genes controlling pattern and growth.

Regulatory Genes

Genes can be divided into structural and regulatory genes. The structural genes are responsible for the similarities in the "housekeeping" proteins to which Wolpert alludes. They are responsible for building the proteins the body is made of. The regulatory genes turn the structural genes on and off, thus affecting the development of the embryo and the organism, as well as the physiology of the organism. M. C. King and A. C. Wilson write:

> Small differences in the timing of activation or in the level of activity of a single gene could in principle influence considerably the systems controlling embryonic development. The organismal differences between chimpanzees and humans would then result chiefly from genetic changes in a few regulatory systems, while amino acid substitutions in general would rarely be a key factor in major adaptive shifts.[16]

As LaFollette and Shanks explain, understanding the role of regulatory genes in evolution is

> . . . crucial to a proper understanding of biological phenomena. First, they focus our attention not merely on structural similarities and differences between organisms but also on the similarities and differences in regulatory mechanisms. Second, they illustrate an important fact about complex, evolved animals systems: *very small differences between them can be of enormous biological significance. Profound differences between species need not indicate any large quantitative genetic differences between them. Instead, even very small differences, allowed to propagate in developmental time, can have dramatic morphological and physiological consequences* [emphasis added].[17]

Even a few examples of "dramatic morphological and physiological consequences" illustrate how the similarities in an organism's structure make it appear at first glance that we can use animal models, while the profound differences in molecular composition demonstrate why the model breaks down upon further examination. A single amino acid difference between humans and nonhuman primates prevents HIV from binding to the same cell receptor in nonhuman primates. A single amino acid difference is responsible for the difference between the hemoglobin molecule in humans with normal blood and the hemoglobin molecule in patients with sickle cell anemia and a single amino acid difference is responsible for cystic fibrosis. That is how very small differences on the cellular and sub-cellular level lead to dramatic differences in the organism as a whole.

LaFollette and Shanks question the relevance of extrapolative data based on these differences:

> Since phylogenetically related species, say mammals, have all evolved from the same ancestral species, we would expect them to be, in some respects, biologically similar. Nonetheless, evolution also leads us to expect important biological differences between species; after all, the species have adapted to different ecological niches. However, Darwin's theory does not tell us how pervasive or significant those differences will be. This again brings the ontological problem of relevance to the fore. Will the similarities between species be pervasive and deep enough to justify extrapolation from animal test subjects to humans? Or will the biological differences be quantitatively or qualitatively substantial enough to make such extrapolation scientifically dubious?[18]

We now know that with systems as complex as the anatomy, physiology, and biochemistry of human and nonhuman animals, even infinitesimal dissimilarities are not incidental. No species is one hundred percent isomorphic—that is, having a one-to-one correspondence between all elements in each system—with another. Therefore, it is impossible to claim that nonhuman animals are completely isomorphic to humans. Since an organism's systems (organs, tissues, and so forth) may differ in subtle and unknown ways, the same exposures often cause different reactions in different species. So, going back to CAM theory, for a CAM to be predictive, "there should be no causally-relevant disanalogies between the model and the thing being modeled."[19] In light of what we know about evolutionary biology, this is impossible without total knowledge of both the model (animal) and the organism being modeled (human). As a result, animal models fail one of the central criteria of sound scientific theory: predictability.

As noted earlier, CAMs must have common causal features, causal connections between the features, and no causally relevant disanalogies. None of these characteristics can be known until we know one hundred percent of everything about the phenomena in both species in question, for example, humans and the experimental animal. Animals can only be proven to be models *empirically*. That is to say, we must know what happens in humans first, then study animals to see if one in particular replicates the human condition. Only by comparing results from experiments on animals with the results from human-based data can we determine if nonhuman animals are sufficiently similar to humans to allow the extrapolation of results. We cannot extrapolate the data *prospectively;* we can only do it *retrospectively.* Animal models can usually be found that replicate human data, but, as we learned in the previous chapter, verifiability is not the same as predictability. In this case it is merely demonstrative and redundant. In *Statistical Science,* scientists Freedman and Zeisel stated:

> Numerical assessments of human risk, even if based on good animal data, seems well beyond the scope of the scientifically possible. . . . The dose-response models now used in numerical extrapolation are quite far removed from biology. . . . In the present state of the art, making quantitative assessments of human risk from animal experiments has little scientific merit. Valid extrapolations would be possible only on the basis of mathematical models grounded in biological reality and carefully tested against empirical data. . . . [20]

In 1999, Michael S. Rand, D.V.M., stated:

> Unfortunately, while the factors that play a role in model selection can be listed, there is no way of generalizing the importance of each factor. Model selection is very much the prerogative of the individual scientist, who therefore is responsible for convincing the rest of the scientific community that her or his choice is valid.[21]

Even the *Handbook of Laboratory Animal Science* acknowledges the lack of predictive value in animal models:

> It is impossible to give reliable general rules for the validity of extrapolation from one species to another. This has to be assessed individually for each experiment and can often only be verified after first trials in the target species.[22]

Occasionally defenders of animal-model research acknowledge that animals fail as CAMs due to their lack of isomorphism, yet continue to insist that animal models are necessary to evaluate a drug or procedure in an *intact* system. It is true that life processes are interrelated. The liver influences the heart, which in turn influences the brain, which in turn influences the kidney, and so on. Thus, the response of an isolated heart cell to a medication does not confirm that the intact human heart will respond as predicted by the isolated heart cell. For example, the liver may metabolize the drug to a new chemical that is toxic to the heart while the original was not. Nor can cell cultures, computer modeling or *in vitro* research replace the living intact system of a human being.

Nevertheless, the real question here is: despite its imperfection, can the animal model do better than the nonanimal methods in terms of predictive value? To determine the answer, let's return to our previous example on causal/functional asymmetry. Let's assume system S_1 has causal mechanisms $\{a,b,c,d,e\}$ and system S_2 has causal mechanisms $\{a,b,c,x,y\}$. If we stimulate the sub-system $\{a,b,c\}$ of S_1 with stimuli s_f and get result r_f, then we would expect to get r_f from $\{a,b,c\}$ of S_2 as well—if the model is viable. However, this outcome will be highly probable if and only if $\{a,b,c\}$ are *causally independent* of $\{d,e\}$ and $\{x,y\}$. In biological systems, as those who argue in favor of intact systems emphasize, almost all systems interact. We have no a priori reason to think otherwise.[23] So the intact systems argument fails.

Let's put it a different way. We have two books, Book A and Book B. Book A has (a) pages, (b) a cover, (c) words, (d) an author, (e) a publisher, and the unknown contents f. Book B has (a) pages, (b) a cover, (c) words, (d) Shakespeare as the author, (e) a publisher and (f) all the works of Shakespeare as the contents. Can we use Book B as a *model* to predict the contents of f in Book A? Obviously not. There must be a strong *causal* connection between $\{a \ldots e\}$ and f in order for causal analogical reasoning to be true—causal connections that animals do not *predictably* demonstrate for humans. Again, the animal-model theory fails to offer predictive value—a key criteria for sound scientific theory. Not all characteristics of the model need to be present in the organism being modeled. But the similarities must be *causally* relevant. In this example, f must be *causally* connected to $\{a \ldots e\}$.

Follow the Money

As we have seen, evolutionary theory provides the foundation for a compelling argument against the efficacy of the animal model as a paradigm for studying human disease. And, as noted earlier, evolutionary theory is the single organizing principle of modern biology. It has replaced the dogma of creationism under which the animal model first appeared. The animal model also fails the test of axiomatic-deductive reasoning. Testing on animals or studying diseases in animals *can* be done in a scientific manner. For example, if a researcher wants to learn about FIV in cats, he can experiment on cats and that knowledge can be extrapolated to other cats. He cannot extrapolate the results to humans with HIV however. When the axiom is "animals and humans have so much in common that we can extrapolate the results from animals to humans," the results will be disastrous—not because the scientific method was not followed, but rather, because the axiom itself is incorrect.

To support the animal-model theory, then, is to deny the validity of evolutionary biology. It would seem that no reasonable scientist would do such a thing, for that would throw all of modern biology into reverse gear. Claude Bernard and other scientists who conducted animal experimentation in the 1800s rejected the theory of evolution and refused to acknowledge the differences that speciation introduced.[24] (For that matter, they also rejected the notion that statistics could be of any benefit in medicine. Bernard stated ". . . if based on statistics, medicine can never be anything but a conjectural science; only by basing itself on experimental determinism can it become a true science . . . I think of this idea as the pivot of experimental medicine, and in this respect experimental physicians take a wholly different point of view from so-called observing physicians."[25]) Bernard's modern-day followers ignore evolutionary truth every time they conduct an animal study for the purpose of learning about human disease.

Yet the use of the animal model remains standard, accepted practice in virtually every sector of biomedical research. Despite the fact that nonanimal modalities, such as epidemiology, *in vitro* research, clinical research, autopsies, mathematical and computer modeling, and other human-based and technology-based research offer results that are much more predictable and reliable, animal-model research continues to thrive. Why?

The answer lies not in science, but rather in the system under which scientific research is conducted in the United States and other countries in the world. It is a system that routinely compromises honest intent, in the relentless pursuit of money, power, prestige, and job security. And, as Irwin Bross, Ph.D., Director of Biostatistics at the Roswell Park Memorial Institute for Cancer Research in Buffalo, New York for 24 years, stated in *Scientific Fraud vs. Scientific Truth* "Establishment is the enemy of enterprise."[26] The establishment—the NIH, the American Medical Association (AMA), and the National Cancer Institutes—perpetuates itself by maintaining the status quo, and will therefore fund projects that meet that objective. In the United States and elsewhere, the scientific process has become a political process.

The National Institutes of Health are the main source of funding for biomedical research in the United States, giving out hundreds of millions of dollars each year to universities, hospitals, and other research institutions. The majority of this money is used to fund animal-modeled research. Those who propose these animal-model experiments have a quid pro quo relationship with the NIH. The NIH gives them money and they, in turn show up for congressional meetings to tell Congress that the NIH is doing all the right things. Meanwhile, scientists under pressure to "publish or perish" know that an animal-model experiment is the most efficient way to generate quick results that can be turned into a journal article. Adding to the mix are journal editors, who decide which of many submissions to publish. These editors rely on "peer review" panels or committees composed of experts in the field to help them decide which papers to publish. Likewise, the NIH relies on these same committees to decide which grant applications to fund. But who are these so-called experts?

We do not really know, because their names are kept confidential. As a result, they do not come under fire for their decisions. Relieved of any personal responsibility, these "experts" can easily fall prey to making a decision based either on emotion or on a vested interest. If Professor Smith, for example, spent his entire career studying the way viruses reproduce in nonhuman primates, and a grant for studying the way SIV binds the white blood cells in monkeys come up for review, it is more likely that he will look more favorably on that study than on one examining how HIV binds to human white blood cells. Having spent his lifetime studying viruses in nonhuman primates, Professor Smith firmly believes that his life's work is important, and his decisions may be based more on that belief than on the value of the research.

It happens more easily than you think, as scientists are caught in a complex web of politics, professional competitiveness, and just plain greed. Ignorance also factors in. A physicist knows very little about medicine, and a physician very little about physics. When the American Medical Association issues a statement, the physicist usually believes it. By the same token, when the American Association for the Advancement of Science issues a statement, the physician tends to believe it. Many scientists with no knowledge of an issue will support the "establishment" position because as good soldiers in their professional community, they believe it their responsibility to defend the party line rigorously.

The danger comes in the compromise of values. A practicing physician, when deciding on how to treat a patient, makes his decision based on science and evidence, not authority. However, the establishment makes decisions based on maintaining the status quo. If you ask a physician if animal models are vital to medical progress, she may say yes. But if you ask whether she uses the data from animal models when deciding on therapy, she will say no. Put that same physician on a committee to examine research protocols based on animal models, and the power of the group dynamic takes over. Having served on a number of committees, we can attest to the fact that rational thought and a commitment to genuine science are often overlooked in the pursuit of self-serving agendas.

Privately, some scientists admit that the reason for the preponderance of animal-model research lies not in science, but in profit—whether corporate, political, or individual. When a leading cancer researcher was asked about the value of using animal models in cancer research, he replied that it was not in finding new cures but in giving oncologists the confidence to try new treatments. It was a sales gimmick.[27] Others have been more public in getting to the heart of the matter. Bross states:

> After the adulation of the cancer establishment by the mass media that has gone on for so many years, it will be hard for the public to realize that the American Cancer Society and the National Cancer Institute are hazardous to its health. They cannot realize that the establishments are basically political or lobby groups that serve their own self interests and not the interests of the practitioners or patients or public.[28]

Rarely does a scientist start out with the intention of engaging in deceptive behavior. But when the result of an animal-based study cause a media frenzy, as it so often does, an individual's ego becomes inextricably linked with the "discovery." When small flaws appear in the data that was supposed to provide a miracle cure, they are ignored. When at last the treatment is tested on humans and fails, the scientist has little choice but to recant earlier claims, which would tarnish his reputation and reveal the inadequacy of animal-based research, or "play along." So they say, "Yes it appears the animal model failed this time but next time it will work."

Science is, theoretically the logical progression of theories based on thought and observation that explain the material world. Science is theoretically, objective. Hence, when a scientific theory is not supported by facts or when a better theory comes along, scientists should change to the better theory or abandon the one with no support. That is what they should do. Most of the time they do, but occasionally a scientists will cling desperately to a theory or idea long after its usefulness has expired. Some scientists cling to their own pet theories either because of ego, because their livelihood depends on it or for other reasons.

When people have a vested interest in a process or product, it is unrealistic to expect them to behave as if they don't. The *New England Journal of Medicine* published a study in 1999 revealing that for-profit dialysis facilities had a higher mortality rate and decreased rates of placement on transplant waiting lists than not-for-profit owned dialysis centers.[29] The reason should be obvious: the longer a person stays in dialysis waiting for a transplant, the more money the dialysis center makes. Dialysis centers and animal experimenters are not unique. According to Dennis Cauchon of *USA Today*, September 25, 2000:

> More than half of the experts hired to advise the government on the safety and effectiveness of medicine have financial relationships with the pharmaceutical companies that will be helped or hurt by their decisions. . . . The experts are supposed to be independent, but *USA Today* found that fifty-four percent of the time, they have a direct financial interest in the drug or topic they are asked to evaluate.

These conflicts include helping a pharmaceutical company develop a medicine, then serving on an FDA advisory committee that judges that drug. . . . These pharmaceutical experts, about 300 on eighteen advisory committees, make decisions that affect the health of millions of Americans and billions of dollars in drug sales. With few exceptions, the FDA follows the committees' advice.

The article also points out that:

At 159 FDA advisory committee meetings from Jan. 1, 1998, through last June 30 [*USA Today*] found: At ninety-two percent of the meetings, at least one member had a financial conflict of interest. At fifty-five percent of meetings, half or more of the FDA advisors had conflicts of interest. Conflicts were most frequent at the fifty-seven meetings when broader issues were discussed: ninety-two percent of members had conflicts. At the 102 meetings dealing with the fate of a specific drug, thirty-three percent of the experts had a financial conflict.

Logic and the Animal Model

Clearly, the vested interests have much to protect, and naturally become defensive when their interests are questioned. The reliability of animal-model research is just one of many examples. Because public debate only serves to bring attention to the fallacies of their argument supporting the use of animal models, the vested-interest groups are reluctant to debate the issue in an open forum. We have appeared for scheduled debates at many universities including the University of Minnesota, the University Nevada, Reno, and many others but were the only debaters there; the opposition mysteriously canceling at the last minute. When unable to avoid a debate, those with a vested interest in animal models offer a number of reasons for justifying animal-model research, none of which can be considered scientifically valid. Let's examine some of them.

In his book *Full House*, Stephen Jay Gould wrote, "nothing can be more misleading than formally correct but limited information drastically yanked out of context." When proponents of animal-model research defend the paradigm, they are doing much the same thing. They frequently cite cases where specific animals and humans gave the same result to a medication, test, or physiology experiment. But, by pointing out specific examples where animal research turned out to have human correlations and deducing from them a pattern of success, they mistake aberrance for a representative sample. In setting out exceptions and saying that they are the rule, animal researchers make the non sequitur called neglect of negative instances—ignoring data that would prove one wrong. (A *non sequitur* is literally a conclusion that does not follow from the premises or evidence upon which it is based.)

Defenders of animal-model research suggest a causal relationship between animal experimentation and all the great discoveries of medical science. They simply state as fact that such advances as the decrease in infant mortality, the discovery of antibiotics and vaccines, the development of artificial joints, and the invention of imaging technology were made possible through animal-model research. However, they do not provide a theory as why such a phenomenon could have happened. Nor can they produce articles from peer-reviewed scientific literature explaining each step of discovery in support of their conclusion. The fact that animal studies were going on at the same time as these great discoveries and inventions implies nothing more than a casual or parallel relationship between animal-model research and medical advances. The animal studies, in many instances, did not cause, nor did they directly lead to, the discoveries.

This is how, in the hierarchy of respectability in science, the argument presented by defenders of animal-model research breaks down. They gloss over the first two tenets of respectability—that the study of and verification by scientific theory is most respected, and that examples from observation and controlled experiments that refute or support the theory are second. Then, they state inflexibly that animal studies have been used to cure disease. Lacking a strong theoretical and evidence-based foundation for their position, animal researchers inevitably turn to the argument that animal models are useful simply because they say they are. Their argument represents classic *fallacious reasoning*:

- We are scientists and researchers who state that animal models in fact do lead to cures for human disease
- Scientists and researchers are truthful and want what is in the public's best interest
- Therefore, when we say that animal models are useful, it is true

Lakatos adamantly rejected the view that truth is whatever the majority believes, just because the majority believes it. For centuries, people believed that the earth was flat, and that disease was caused by the presence of evil spirits. But the fact that these beliefs became part of the collective knowledge did not make them correct.[30] One reason society believes animals make good models for human disease is the fact that they hear it so often. A twice-told tale is believed more strongly than one told only once and the vested-interest groups makes sure their tale is told several times each day from numerous sources. For example, in the United Kingdom it is not uncommon for the Research Defense Society to spend hundreds of thousands of dollars on propaganda supporting the animal model. Peter Medawar said in *Advice to a Young Scientist*:

I cannot give any scientist of any age better advice than this: the intensity of the conviction that a hypothesis is true has no bearing on whether it is true or not. The importance of the strength of our conviction is only to provide a proportionately strong incentive to find out if the hypothesis will stand up to critical evaluation.

Researchers using the animal model also employ fallacious reasoning in the form of *ignoratio elenchi,* or *irrelevant conclusion fallacy*—drawing a conclusion that says nothing about the premise it is supposed to be supporting or contradicting. In this form of reasoning, proponents of animal-model research focus on issues that no one can argue with, such as the need for better prenatal care, vaccines for young children, good health, and so forth, rather than focusing on the central issue, which is how animal models are used to accomplish these goals. They say, "Children need to be cured of cancer therefore animal experiments should continue." Granted childhood cancer is terrible and cures need to be found, but what does it have to do with whether animal models are useful?

Another similar, popular tactic is the *ad populum* fallacy, which appeals to emotion rather than reason. Advocates of animal models appeal to mass sentiment to support a conclusion for an argument that cannot be supported by the facts. They ask, "Do you want to see sick children suffer when research on animals could cure them?" Most people would reply, "Of course not." But the fallacy is that animal models can't cure them. Their posters and other promotional material often show happy, healthy children at play, while the text boasts that they were cured of some deadly disease because of research on animal models. Statements totally unsubstantiated from the medical literature.

Because public debate serves only to bring attention to the fallacies of their argument, supporting the use of animal models, the vested-interest groups are reluctant to debate the issue in an open forum. However, they do not hesitate to disseminate all manner of propaganda in defense of their position, even to the point of publishing a workbook for children grades four through eight fabricating that animals are needed in biomedical research.[31] Books and articles admitting the suffering of animals in research, while supporting the validity of animal models, send a message the establishment loves to hear: that animal research is a "necessary evil." They know that as long as the animal model is viewed as a necessary evil, and that, try as they might, some animals must suffer so people can live longer, healthier lives, their jobs are quite safe.

At the same time, those who challenge the dogma of animal-model research are repeatedly denied influence over consensus when animal experimenters fail to participate in prearranged debates or allow the publication of articles on the subject. As long as the vested-interest groups control who is and is not allowed to speak on an issue, only one view will be heard.

By denying the inability of animal models to predict human outcomes, practitioners of animal-model research fail completely in living up to their responsibility as the "community of advocates." The community of advocates responsibility mandates them to attempt to solve the problems their theory faces, explain the anomalies, compare the success of their theory to the success of competing theories, and try to prove or disprove their theory. We have seen how the animal model community makes little attempt to develop their theory toward solutions, shows no concern for attempts to evaluate the theory in relation to others, and is highly selective in considering confirmations and disconfirmations. Because they have a vested interest in the continuation of animal-model research, they have made no effort to learn new scientific research methods that would replace the animal model.

Conclusion

As Thagard said, a theory is pseudoscientific if it has been less progressive than alternative theories over a long period of time. The animal model paradigm appeared viable in the 1800s because at the time we knew so little about human anatomy and physiology. On the gross macroscopic level, all animals were alike. Dogs had hearts; so did humans. Cats had electrical activity in their brains; so did humans. In the past, animal studies provided the correct answer to questions about the very big picture of how a living organism functions. But today, scientists are studying phenomena on the very level that differentiates one species from another—the cellular and molecular level. Just as modern physics replaced classic physics without destroying Newton's laws of motion and gravitation, so modern, evolutionary-based biomedical research can render the animal model obsolete without saying that animals and humans are completely different.

Classic Newtonian physics explained many things about the nature of motion and energy. But it was what Newtonian physics did *not* explain that led to modern physics. Newton's laws were one of the most important discoveries ever in part because they revealed that ghosts and demigods who could capriciously change the rules of nature on a whim did not rule nature. Newton showed that nature is ruled by law. Before Einstein, scientists were frustrated because everyone knew Newton's laws were correct but James Clark Maxwell's discoveries about light and electromagnetism seemed to contradict them. Everyone was looking for reasons that would prove Maxwell wrong—everyone except Einstein. He suggested that maybe Newton was wrong. This was considered heresy in the early 1900s. (Just as suggesting that animal models may be harming humans is considered heresy by the vested-interest groups today.) But Einstein proved that Newton was in fact wrong about this area of physics. Quantum mechanics and relativity theory has been proven as thoroughly as anything in science. Without it we would not have lasers, transistors, electron microscopes, and many other technologies we take for granted. Newtonian physics could not have given us these things.

Newton's laws can be used to put a man on the moon and explain or predict most of what we encounter in everyday life. Yet, Einstein's theories better explain what happens at speeds approaching the speed of light or when gravity is much different from here on earth. Such conditions occur in black holes or occurred in the first milliseconds of the Big Bang. Einstein showed that space and time are relative; they can be changed depending on your velocity and gravitational field. Modern physics also explains the more mundane. So it is better than classic, Newtonian physics.

In vitro research, epidemiology, autopsies and other human-based or technology-based research modalities can be used to explain what we learned from animal models many years ago and they allow us to discover things animal models do not. Hence animal models have outlived their usefulness. Yes, they can still be used to show a high school biology student that nerves and blood vessels course through our bodies. One can also use cadavers, computer programs and other aids, to demonstrate that the same principles apply to all mammals. But in modern-day biomedical research we are far beyond that level. Animal models are limited in their scope, and their scope peaked 100 years ago. In *Voodoo Science*, Robert Park states:

> The success and credibility of science are anchored in the willingness of scientists to obey two rules: 1. Expose new ideas and results to independent testing and replication by other scientists. 2. Abandon or modify accepted facts or theories in the light of more complete or reliable experimental evidence.[32]

In this age of technological advancement, we have reached the point in our knowledge of biology where animal models are no longer of value and will in all likelihood mislead us. Therefore, this archaic paradigm must be replaced if we expect to pursue scientific knowledge in accordance with the rigorous standards that ensure scientific integrity, advance scientific knowledge, and improve the quality of human life. Having shown how the animal-model fails as a scientific paradigm from a *theoretical* viewpoint, we can now turn our attention to how the use of animal models fails in *application*. In the following chapters, you will see how using animal models in predicting human outcomes of disease delays medical progress by providing false and misleading data—and has actually cost people their lives. We will also show where the great discoveries of medicine really did come from and where they are coming from today.

chapter 3

Genes, Technology, and Internal Medicine

Medicine is learned at the bedside and not the classroom.

—Sir William Osler

All medicine revolves around internal medicine. This chapter examines the revolutionary role of genetic science on medicine. As such, it is a forum for illnesses not addressed elsewhere. Likewise, the chapter includes fresh information on cancer that has emerged since publication of our previous book *Sacred Cows and Golden Geese.*

Internal medicine is one of the broadest of the medical specialties. Internists address diseases affecting everything from respiration to micturition. It is the foundation of all other medical subspecialties. Indeed, most of these specialists begin their residencies with three years of internal medicine before even embarking on their chosen fields. As a discipline, internal medicine provides the fundamentals of medical training. During the medical student's rotation in internal medicine, he or she learns how to take a history, conduct a physical exam, enter data in a chart and perform other functions that fill every physician's day. But of all medical students who rotate through internal medicine only a few—the internists—choose to specialize in internal medicine. Patients rely on their internists to guide and monitor their health, and may visit them for conditions as disparate as constipation and intracranial aneurysm. If the infirmity cannot be assuaged, the internist refers the patient to another specialist—a cardiologist, urologist, oncologist, a surgeon, and so forth—a physician who began his or her medical training just as the internist did, by studying internal medicine in their days as a medical student. Therefore, in education and practice, internal medicine is a touchstone for overall health.

Health occurs where our genetic predisposition, our physiological functions, and our lifestyle intersect. When all these are optimal, our health is good. When it falters, it means one or more of the variables are askew, sometimes only slightly. And, if not rectified, the faltering aspect will diminish our well being in other regards. Viewed this way, one can easily imagine the challenge in trying to re-create human holism from a nonhuman. This chapter will explore that challenge in more detail.

The Lesson of the Microcosm

Our bodies are intricate composites; with each organ's operation depending on the others in ways science is still bringing to light. The more we learn, the more new puzzles present themselves. Though we tend to think more frequently of the *outer* universe as an infinite, unquantifiable space, it turns out that our *inner* universe, the microcosm, is much the same. The smallest piece of our bodies is only the smallest *known* piece. There is always more to discover about its details and the details on those details. Thus far, exploration into the microcosm confirms that it too is infinite and unquantifiable.

Closer to our subject, the folly of human hubris in this regard is everywhere evident in the history of animal models. Take, for example, the subject of genes. Researchers who use animal models always use the logic of our close kinship with apes, in particular, to justify their work. This was true even before the monk Gregor Mendel's research began to prove the basis for genetic proximity. (Mendel, experimenting on pea plants in the 1860s, was the first to describe patterns of inheritance, thus laying the groundwork for the discovery of the gene, the unit of inheritance.) As genetic research continued after Mendel, the rationale for experimenting on nonhuman primates seemed even stronger. We came to understand that humans and nonhuman primates have up to ninety-nine percent of their genes in common. That was a huge similarity, by contrast with other species—say the fifty percent of the genes we have in common with bananas—or so we thought. One percent did not seem like a very great gulf. The commonality, naturally, appeared to reinforce the reliability of using nonhuman primates as surrogate-humans for research.

However, looking more closely at genes, scientists discovered that they are complicated to an extent that widens the gulf between species. In the previous chapter, we described DNA, the building block of all life and its four different nucleotide bases—adenine (A), guanine (G), thymine (T), and cytosine (C). In the human genome alone, there exist 3.2 billion DNA base pairs—GC and AT combinations. There is a lot of potential for variability there. These three billion base pairs are segregated into 40,000 or more genes, and each *gene* is a code or recipe for a specific protein made of amino acids. The individual proteins may be from as few as several amino acids in length to as many as hundreds, which we shall discuss in more detail in the Development of Medications chapter.

So, whereas all plant and animal species share much of the same genetic material, it is the vast *composition,* or arrangement, and switching on and off of this genetic material that makes all the difference. Given that new information about these compositions arises daily, it is not a stretch to say that the differences seem infinitely intricate. Therefore, what is important to accept here is that the one percent difference between, say, your brother-in-law and a bonobo is actually a dissimilarity of greater magnitude than anticipated. The discovery process indicates that the microcosm is an onion with more and more internal layers. There is no end to it. Very small differences on the cellular and subcellular levels translate into huge differences on the gross or macroscopic level.

It is not just between species, but also within species that small differences present huge discrepancies. That makes the legacy of unraveling the human genome priceless in its ramifications. The human genome is an instruction manual written in a code that, to date, only cells have been able to read and respond to. We know that there are several billion letters in the human DNA genome grouped together into about 40,000 genes, all found in the cell nucleus. And each gene has, approximately, between 800 and 2.7 million base pairs. However, slowly and bit by bit, scientists are beginning to fathom the genome's directives.

As scientists continue to decode this instruction manual using *in vitro* research and epidemiology (epidemiology is a branch of medical science that deals with the incidence, distribution, and control of disease in a population, epidemiologists study human populations in order to learn about disease), they learn much more about how disease happens and why it happens to some humans and not others. The ability to match specific genes to specific diseases makes genetic medicine one of the most promising fields in medicine today.

The first task is to isolate the *genetic markers* for these diseases. For instance, scientists identified five genes responsible for a hereditary spastic paraplegia.[1] These five genes are that disease's genetic markers. As examples of more markers, scientists now know that recurrent deep vein thrombosis is twice as high among carriers of mutations (altered DNA) in both the factor V and the prothrombin gene than those who have one or neither mutations.[2] Papillon-Lefevre syndrome is an autosomal recessive disorder (autosomes are chromosomes other than sex chromosomes), characterized by severe early onset periodontitis, the loss of supportive bone around the teeth. Cathepsin C mutations predispose people to this affliction.[3]

Differences between genotype and phenotype (the physical characteristics of the person) perturb the search for genetic markers. Not all genes will be expressed, and this decreases predictability. Given the complications, just within a single species at a time, does it not make more sense to study and attempt to cure the species in question—Homo sapiens—particularly as the methods of studying and curing are more and more comprehensive?

In some cases, knowing the genetic markers for a disease provides physicians with an early warning system so they can give patients health counseling, hopefully before predisposition to the disease leads to disease itself. Take celiac disease, for example, a chronic intolerance to gluten. Since celiac disease is hereditary, physicians can screen relatives in a family with the propensity for the disease before clinical signs appear. The markers also create a more comprehensive picture of the prognosis. This information will become even more valuable as science perfects ways for redressing nature's mistakes through genetically determined therapies.

The Case of Single Gene Defect Diseases

Most human diseases result from multiple gene disorders, but more than three thousand are caused by single gene defects, one little gene among those 40,000. Two of the most common are cystic fibrosis and sickle cell anemia both resulting because a single amino acid is out of position in one process-supporting protein.

Single nucleotide polymorphisms, so-called "SNPs," (pronounced *snips*), are places in any individual's genetic code where the DNA deviates by a single nucleotide, the substitution for T by a C or G or A. Single nucleotide polymorphisms occur once in every 500 to 1000 base pairs (that is one-tenth of a percent of the time). Even though this sounds like a little, it amounts to dozens of times in each of the estimated 40,000 genes. SNPs can introduce susceptibility to specific disease such as with cystic fibrosis and sickle cell anemia. Even though scientists have already identified many human genes related to disease, researchers who use animal models continue to look for parity between animals and humans in these single gene defects. However, as the medical textbook *Nonhuman Primates in Biomedical Research*, states:

> Despite the enormous number of human diseases caused by single gene defects, a 1986 review of the literature revealed no instances in which research with nonhuman primates had revealed a well-defined hereditary disease controlled by a single gene.[4]

Yes, we do have genes in common with animals. But this textbook admits that when animals have single gene defects, the defects do not translate to humans. Say a certain chimp is missing a gene on a specific chromosome. When we look for the corresponding gene defect in humans with a similar disease, it is not there. That is what this means.

The mapping of the human genome had been more or less completed for a few months when on February 12, 2001, the biotech company Celera announced the completion of the mouse genome. Mice have the same general layout as humans and genes that exist in humans also exist in mice. But can we use the data from the mouse genome to make predictions about human response to disease? No. Despite the lack of correlation, genetic engineers reasoned that deliberately introducing a particular SNP, or defect, into another species might result in animal models for exploring these diseases and creating pharmaceuticals. By plugging a few generations' worth of rodents with mutated human genes, they indeed managed to concoct gene-defective animals. Genetically engineered mice, also called *transgenic* mice, seemed as if they would be the most economical models, especially as they continue to reproduce like mice, spawning generation after generation of merchandisable little lab animals.

Unfortunately, though successful at inserting the single gene defect, scientists could not control variables outside that one errant gene, such as the rest of the mouse genome, its habits and so forth. These "incidentals" regularly put research off-kilter. Therefore, scientists are not yet able to deal with the extraneous stimuli that direct the outcomes of experiments in still unpredictable ways. According to a publication devoted to examining the efficacy of animal models:

> ... specific genetic defects can be as difficult to identify and characterize as those of their human counterparts; and affected animals often differ from unaffected controls [normal animals] in the genetic factors additional to the gene in question. ... There are several limitations in relation to the usefulness of the current approaches to developing transgenic disease models, particularly since many diseases are multifactorial. Problems persist when extrapolating data obtained by using such transgenic animals to the disease condition in humans.[5]

That is to say, apart from the gene in question, transgenic animals have a full array of other genes, organs, and systems that interact with the gene. Importantly, many diseases appear to result, not just from the gene defect, but also from these other factors, internal and external. Extrapolation of results to humans has no way of trafficking these differences, as you will read later in this book. The sections on sickle cell anemia and on cystic fibrosis in the Pediatrics Chapter include particularly persuasive examples of the ineptitude of the transgenic animals. Despite this, the U.S. government (via the National Heart, Lung and

Blood Institute and the National Human Genome Research Institute) and private institutions have spent over one hundred million dollars to sequence the rat genome in hopes of learning more about humans.[6]

The media recently heralded as a scientific triumph the work of scientists at the Oregon Health Sciences University (OHSU) in creating "ANDi" (inserted DNA spelled backward), a rhesus monkey that had a gene from a jellyfish inserted into him.[7] OHSU touted this as a great medical breakthrough because they claimed that genetically altered non-human primates can now be utilized for investigating many genetic disorders such as muscular dystrophy and cancer as well as studying diabetes, Alzheimer's, breast cancer and HIV. Why so? They have no proof of this. All the OHSU researchers did was simply apply to monkeys what they been practicing on mice for almost thirty years. They have had absolutely no success in extrapolating data for human diseases, neither with mice or monkey. This 'medical breakthrough' has no meaning for people suffering from diseases.

Adding one gene to a nonhuman primate is not going to make a man out of a monkey. It does result in increased publicity for OHSU, and will translate into funding, but cures and treatments for human disease will not follow. The resources that went into creating ANDi should have gone to research that may have actually resulted in treatments and cures; clinical research, test tube research, basic science research, and the myriad other reliable research methods. Even the editors of *New Scientist* stated, "There's no doubt the research is symbolic. We can now give an extra gene to a primate. But claims that it is anything more than that are monkeyshine."[8]

Recombinant Technology and Gene Therapy

How did they get the extra gene into the primate? Genetic engineers have learned to perform a sort of genetic surgery wherein, through cut and paste, they create new DNA for experimental or therapeutic objectives. The first task is to identify the genes (be they mutated or healthy) in question, then pluck them off chromosomes using *restriction enzymes*. Restriction enzymes, which come from bacteria, are like genetic clippers. They can recognize a particular sequence in the chain of chemical units—the nucleotide bases that make up the DNA molecule—then cut the DNA at that location. Next, genetic engineers splice DNA fragments into an existing DNA sequence using an enzyme called *ligase* that effectively glues it into a viable bond. The result is *recombinant DNA*.

Requiring a way to copy or "clone" that novel piece of DNA, genetic engineers developed a sort of highly miniaturized Trojan horse within a Trojan horse. This is an infiltration and proliferation scheme for duplicating recombinant DNA in large quantities. Bacteria, being single-cell organisms, reproduce very fast, and any DNA within them multiplies in volume exponentially, so they can be hijacked and used as DNA factories. But a bacterium will not welcome just any DNA. Therefore, a transport, known as a *vector* is necessary. This is a form of DNA that the bacterium recognizes as acceptable, such as certain viruses small loops of bacterial DNA called *plasmids*. The recombinant DNA in bacterial clothing passes into the bacteria and soon burgeons into many identical double helices.

In addition to its being used for creating whole new organisms such as transgenic animals, recombinant technology has been invaluable for the production of human body chemicals—the blood-clotting agent missing in hemophiliacs and insulin for diabetics among them. This utterly marvelous technology is light years more sophisticated and useful than any animal model. The human insulin grown in large quantities within *E. coli* cells results in an abundant source of insulin. Scientists are still pushing for improvement using genetically modified human hepatocytes—liver cells—to produce and secrete human insulin. The insulin is not yet sensitive to plasma glucose levels, but further engineering may provide better insulin at less expense than the *E. coli* used previously.[9]

Medical experts anticipate that scientists will perfect methods of inserting healthy genes into human patients suffering as a result of missing or mutated genes. This form of treatment is called gene therapy. Gene therapy, thus far, comes in two varieties—germ-line gene therapy and somatic-cell therapy. Germ-line gene therapy alters germ cells—sperm or eggs—resulting in permanent genetic change within the whole emerging organism and subsequent generations. Plot to the movie *GATTACA* aside, germ-line gene therapy is not yet an option for humans, though genetic engineers regularly use it to breed transgenic animals. It is the far less controversial somatic-cell therapy that seems to offer so much hope as a remedial measure for genetic-based diseases. Scientists add therapeutic genes to unhealthy tissue, hoping to supply functional genes to cells lacking that function.

As we have written, genes cannot, by themselves, permeate the cell surface to supply the needed changes to the DNA. They require a vector such as a virus. For this reason, scientists have been using a genetically modified respiratory virus, called an adenovirus, to deliver high doses of therapeutic genes. The adenovirus vector effectively buses in the missing gene, delivering it right to the DNA. With a healthier double helix in place, the hope is that the cells will replicate to include the imported gene.

Naturally, government agencies are attentive to progress in gene therapy, as it is a significant departure from conventional therapies. The procedure suffered a substantial setback in September 1999, due in part to animal-modeled research. Eighteen-year old Jesse Gelsinger suffered from a rare liver disease known to originate on a genetic level. Gelsinger's faulty gene resulted in a deficiency of the enzyme ornithine transcarbamylase. Research teams had produced volumes of the gene and intended to use an adenovirus as a vector for transferring the gene to human patients. In preliminary experiments, they injected the missing gene, via the adenovirus, into animals, which responded well to the method. Based on this work, physicians at the University of Pennsylvania then injected Gelsinger. The adenovirus, tragically, multiplied out of control and killed him.[10,11] In a review of the tragedy, the journal *Science* stated:

> Animal trials had indicated a higher transduction rate [meaning that the gene was accepted into the cell], but the Penn team doesn't know why the adenovirus works less efficiently in humans.[12]

As usual, animal models misled research. Scientists may have been mistaken by studies in mice because of a difference between the mouse liver and the human liver. *Science* stated:

> Animal data may have given clinicians false hope that adenovirus would work well in the human liver. A key docking site adenovirus uses to enter a cell, known as the Coxsackie adenovirus receptor (CAR), is much more abundant in mouse livers than in human livers. In fact 'rodent models might be misleading' for gene therapy, says Jeffrey Bergelson of the Children's Hospital of Philadelphia. Bergelson published a paper in 1998, a year after the trial began, reporting that he found 'barely detectable' signs of CAR in human liver, while signs of CAR were 'off the chart' in mouse liver. One implication, Bergelson notes, is that clinicians relying on the mouse model may find it necessary 'to give higher and higher doses' of the vector to deliver genes to the human liver.[13]

University of Pennsylvania researchers initially claimed that nothing from the animal models had predicted Gelsinger's fatal outcome. It was then learned that two monkeys had died from similar experimental gene therapy. Playing both sides of the controversy, the animal-experimentation industry jumped on this as proof that animal models had predicted the fatal outcome. The journal *Nature* pointed out that the two monkeys who died were models for a different disease and received a different therapy.[14] Many other animals had similar gene therapy and lived.[15] Importantly, if performing the same experiment on many different animals produces many different results, the experiment is simply not predictive.

After Gelsinger's death, it came to light that Penn had neglected to report to the FDA over six hundred untoward reactions to gene-therapy in other humans. Penn failed to stop the gene therapy study that killed Jesse Gelsinger even though previous patients had incurred grade III toxicity from the gene treatment.[16] These reports were not mandated by laws or regulations so most researchers did not report them. If they had, the problem with the vector might have been recognized and Gelsinger's death prevented.[17]

Because there are fewer Coxsackie adenovirus receptors in humans than in mice, physicians had to inject Gelsinger with overabundant virus. His system could not handle it and this is why he died. To claim that extrapolation from animal models did not contribute greatly to Gelsinger's demise simply is not true. The physician responsible for administering the gene therapy to Jesse Gelsinger, James Wilson, has now done further animal experimentation and has reproduced in mice and monkeys what happened to Gelsinger.[18] Once again, this merely demonstrative data is being heralded as a breakthrough.

Henry I. Miller, MD, Senior Research Fellow at the Hoover Institute, was an FDA official from 1979 to 1994 and a member of the NIH Recombinant DNA Advisory Committee from 1980 to 1993. Miller questioned the importance of animal experiments conducted prior to the human clinical trials. He said:

> The results of animal studies, especially those that use a much higher dose than would be administered to humans, is seldom mentioned in the patient consent form; and the fact that 17 OTC-deficient patients had been treated in the University of Pennsylvania trials before Gelsinger without problems also argues against the importance of the monkey data.[19]

To put the research in perspective, Miller recalls when bone marrow transplants were first started in the United States:

> When I was a medical student in the 1970s, for example, bone marrow transplantation was highly experimental. It was performed at only a handful of medical centers in the United States, and success rates were abysmal. But *clinical research* has refined the technique and identified diseases for which the technique is useful: In a genetically determined disease of the red blood cells called thalassemia major, for example more than eighty percent of the patients are now cured [emphasis added].[20]

Undoubtedly more needs to be known about gene transfer therapy, but relying on animal models will not prevent further tragedies and it may cause them. Only human trials based on good human-based research and technology are predictive for humans. Only human trials have led to successful gene therapy. Many people are familiar with "bubble babies" because they have seen these unfortunate children on television. These infants must spend their entire life in a germ-free bubble because they have Severe Combined Immunodeficiency (SCID), for example, no immune system to fight off disease. In 1999, French researchers successfully transferred the gene missing in SCID to two infants suffering from the disease. This was a human breakthrough made possible by *in vitro* research, clinical experience and technology. Other successful gene therapy venues include some forms of cancer and blood disease.

Stem Cells

Growing disenchantment with animal models has led the vanguard to cast aside intact whole organisms in favor of stem cells. Stem cell technology began with *in vitro* research tracking specific cells derived from plants, animals and humans, for longevity in a cell culture medium. They found that some cell types die quickly while others essentially live forever. Tumor cells from humans, for instance, can survive in a culture medium as long as a nutrient source is available. The younger the cell source, the greater the cell longevity.

Stem cells are the youngest of all, "master cells" that can grow into virtually any of the body's cell types. The reason they work is that during embryogenesis, humans develop from a single undifferentiated cell, called a *totipotent stem cell*. Stem cells are unlike any specific adult cell; however, they have the ability to form any adult cell. Stem cells can proliferate in culture, so potentially they provide an unlimited source of clinically important adult cells, if scientists can direct their development in culture into bone, muscle, liver or blood cells. Therefore, stem cells are theoretically, and in some cases actually, an effective source of transplantation material. If the tissue is from humans, the risk of unknown viral infection, such as those that plague xenotransplantation (transplanting organs and cells from animals to humans) decreases. New research suggests that stem cells may eliminate barriers to successful bone-marrow transplants, as an example.[21]

Originally, researchers harvested healthy cells from early stage human embryos derived from *in vitro* fertilized embryos of less than a week old. The embryos were left over after couples had received successful implantations, and the cells were harvested with the full written consent of these clients. Now, these cells can be lab-grown. Other sources of stem cells are umbilical cords and placentas. Physicians can transplant stem cells yielded from this material into children suffering from leukemia, and with fewer problems in regard to matching antigens than with bone marrow transplants. Children with sickle cell anemia, immunodeficiency syndromes, and inherited enzyme deficiencies can also benefit from transplantation of stem cells garnered from umbilical cords and placentas.[22]

Researchers have found that the use of blood stem cells for transplantation significantly increases the tempo of cell production compared with marrow, particularly promising news for patients with high-risk blood cancers.[23] Patients with a broad range of cell-based diseases like juvenile onset diabetes mellitus and Parkinson's disease can also benefit. Replacing faulty cells with healthy ones offers hope for treatment and possibly cures. Likewise, injecting healthy cells to replace damaged or diseased cells could in theory rejuvenate failing organs. Already people with autoimmune diseases—multiple sclerosis, scleroderma, juvenile arthritis, systemic lupus erythematosus, and vasculitis/cryoglobulinemia—have been successfully treated using stem cell therapy. (An autoimmune disease is a state in which the body's own immune system acts against itself.) Around two-thirds stabilize or improve.[24] Stem cell transplantation has been particularly successful in treatment of persistent systemic lupus erythematosus when combined with chemotherapy.[25]

Drug developers can test potential new medications on populations of specific cell types and measure the response. Stem cell technology would permit the rapid screening of hundreds of thousands of chemicals that are presently tested through more protracted means. Stem cell technology has advanced rapidly, and functioning cells from diverse organs and tissue types are growing successfully in labs across the United States. Theoretically, these cell types may mature into functional organs, hence finding ways to direct human embryonic stem cells to become specific cells of clinical importance is key. Instead of funding animal-modeled research, we should be funding research to identify the chemical and molecular pathways that allow stem cells to differentiate into other cells. We should be looking for effective ways to combine gene therapy and stem cell therapy.[26] It goes without saying that animal model-based methodologies seem all the more archaic considering the progress scientists have made in the use of stem cells. These very exciting developments offer unlimited opportunities, which may extend into every field of medicine.

Animal Model Relevance to Cancer and Diseases of the Blood

As you can see, given the complexity of our bodies, justifications for comparison centering on shared DNA are simply not viable. Dissimilarities between humans and other animals, on the molecular level, may give rise to additional differences in a nonlinear fashion. This means that any misinterpretations that arise from animal-modeled data cannot be cut off at the pass, as they would be if they fell as dominoes fall. Since they occur randomly, as in "one rotten apple spoils the whole bushel," there is no telling when, where, and how the error will occur. The errors always undermine the criteria that would make animal experimentation truly scientific (the verifiability, predictability, and so forth that we discussed in earlier chapters). As a result, the animal model is *always* pseudoscientific. This is true whether researchers are crafting medications or studying diseases. The remainder of this chapter examines a few diseases, not addressed in other chapters, which attract research dollars that are too often squandered on the animal model.

Blood Diseases

Our circulatory system is like a freeway and blood is the conveyance in which everything courses through our bodies, everything from vital substances to toxins. In it, red blood cells carry oxygen. White blood cells, as mobile units of the immune system, police deleterious invaders. Platelets help blood coagulate as needed. These components drift in a liquid medium called plasma. When this composition is abnormal, disorders of the blood arise. One such disorder, anemia, occurs when blood is hemoglobin-deficient, that is under-supplied in red blood cells. This fact emerged from epidemiology and *in vitro* research. Anemia has several causes. Blood loss, toxins, or an antibody to the red blood cells can cause anemia, so can iron or B_{12} deficiencies, or functional failure of the bone marrow. One group of anemias is hereditary, and includes sickle cell anemia.

Blood disease has been much elucidated by physicians and scientists who experimented upon themselves. Physicians suspected that diet could influence anemia as early as the nineteenth century, but could not yet confirm it. To determine how gastric secretions affected the red blood cell count Dr. William B. Castle began a novel human experiment in 1926. He removed partially digested food from his own stomach and inserted it, using a nasogastric tube, into patients suffering from pernicious anemia. They got better. Their red blood cell count increased. Castle proposed that two factors were necessary for the body to manufacture red blood cells. One was in food, external to the body. He named this the *extrinsic factor*. The other, the *intrinsic factor,* was within the body in the stomach juices. This experiment linked the stomach to red cell production in the bone marrow. Two decades later scientists identified B_{12}, the extrinsic factor. Another self-experimenter, Dr. Victor D. Herbert, proved the influence of folic acid on red blood cell production in 1961. His red blood cell count decreased when he deprived himself of folic acid. His research linked folic acid deficiency to this form of anemia, which is found throughout the world.[27]

Hemophilia B is a disorder of the blood in which the affected person bleeds easily and profusely. Since a single gene causes hemophilia B, producing deficiency of a blood-clotting factor called factor IX, hemophilia B patients would seem to be candidates for the kind gene therapy described a few paragraphs ago. Animal models suggested that patients who only needed low doses of factor IX supplementation would not benefit from gene therapy. But when researchers gave the gene therapy to them they found the animal models were wrong. The humans, who initially needed only low doses of factor IX, who received the gene therapy were actually requiring *even less* factor IX supplementation to prevent spontaneous bleeding.[28] Though it is too early to say with certainty, gene therapy looks beneficial to hemophilia B patients.

Idiopathic thrombocytopenic purpura leads to blood platelet depletion. (ITP is a condition that arises spontaneously hence idiopathic and which leads to a decrease in the number of platelets hence thrombocytopenic and which in turn leads to bruising hence purpura) ITP patients have huge bruises all over their body. They can bleed internally, even into their brains, which induces strokes. In the 1950s, ITP was a mystery. Then Dr. William Harrington proved, by injecting blood from one of his ITP patients into his own body, that in ITP patients the bone marrow functions properly but platelets are destroyed in the blood stream. Bone marrow biopsies of the time confirmed that the marrow functioned normally. This research also proved that the life span of platelets, hitherto thought to be five days, was actually eight to nine days. Harrington's work offered conclusive proof that the body can attack itself. Disorders such as these are called autoimmune diseases. Although Harrington did not win the Nobel Prize, his research inspired Jean Dausset to experiment on himself. Dausset won the Nobel Prize for immunology in 1980.[29]

As Harrington's study surmised, the enemy of good health is not always external. Our own immune system can create antibodies against itself, resulting in over fifty different autoimmune diseases, which affect women by a two-to-one margin. They include rheumatoid arthritis, multiple sclerosis, juvenile diabetes, cardiomyopathy, anti-phospholipid syndromes, Guillain-Barré, Crohn's disease, Grave's disease, Sjögren's disease, alopecia aerata, myasthenia gravis, lupus and psoriasis, among others. In trying to understand autoimmune disease, scientists have not found the animal model constructive, and often commented that "there are no perfect nonhuman animal models for the study of rheumatic diseases."[30]

In 1965 human observation and *in vitro* follow-up showed that Sjögren's syndrome—in which mucous surfaces, such as eyes and mouth, are very dry—is an autoimmune disorder.[31] Subsequent observation and *in vitro* research determined the molecular mechanism of Sjögren's and other autoimmune disorders.[32] After the fact, animal experimenters found that animals could produce autoantibodies too.[33] A leading immunology textbook states:

> Most investigators would agree that the final proof of human relevance of findings initially made in other mammalian species requires the direct examination of the human system.[34]

Yes, and when the initial findings are in humans, the repetition of those findings in other mammals is unnecessary and resource-wasteful.

Autopsy and *in vitro* studies revealed that autoimmune heart disease occurred after exposure to the bacterium α-hemolytic streptococcus.[35] Human studies also suggested that many autoimmune diseases occur after various viral infections.[36]

Leukemia is a disease of the blood as well as a cancer. Leukemia is a cancer of the blood-forming tissues. Affecting about 30,000 Americans a year, it is a popular research venue. Leukemic cells inhibit normal cell growth and can also lead to a deficiency of red blood cells and platelets. There exist four forms of leukemia, of which acute lymphocytic leukemia and acute myelogenous leukemia are the most prevalent. In 1955, Dr. Thomas Brittingham transfused blood from leukemia patients into his own body. Afterward, his analytical tests demonstrated that cancer was not contagious. (Brittingham also proved that white blood cells have antigens. These, like A and B antigens on red blood cells, can cause transfusion reactions.)[37]

Remember that folic acid affects red blood cell production. As it happens, an estimated one-third of the population has a gene mutation, a SNP, for the enzyme that metabolizes folic acid. This particular version of the gene appears to protect adults against acute lymphocytic leukemia but leaves them vulnerable to acute myelogenous leukemia. However, as is true in most cases, finding a variant gene whose presence is statistically correlated with a disease is not always the same thing as establishing that the variant gene causes the disease. Nutritional, lifestyle and environmental influences impact the susceptibility, as is expressed in an editorial that accompanied the study:

> Single nucleotide polymorphisms (SNPs or single nucleotide changes) provide a powerful molecular tool for investigating the role of nutrition in human health and disease. Their integration into clinical, metabolic, and epidemiologic studies can contribute enormously to the definition of optimal diets.[38]

What this emphasizes is that, in most cases, our genes are not an automatic ticket to health or to illness. Even those who do not have this mutation need to make sure their diet has enough folic acid in it, in order to avoid susceptibility to acute lymphocytic leukemia. Another gene mutation, which affects the transcription factor *CBFA2* (transcription factors "transcribe" DNA into RNA, ready to be "translated" into proteins), causes a familiar platelet disorder that increases susceptibility to acute myelogenous leukemia.[39]

Some forms of leukemia result from exposure to radiation and certain chemicals, especially benzene. Epidemiology also pinpointed high incidence of acute myelogenous leukemia in jet pilots, perhaps due to exposure to cosmic radiation.[40] However, not everyone who comes down with leukemia was exposed to radiation. So screening for leukemia based on the history of exposures would miss these people. Screening for leukemia based on genetic changes would catch it. This advantage is one of genetic medicine's enormous boons.

Traditional leukemia therapies include chemotherapy, radiation, and bone marrow transplants. New treatments include stem cell transplants from the patient or a donor.[41] Recent human studies show that patients who receive stem cell transplants have a twenty-five percent survival advantage over those with the traditional modes. Alternatively, physicians can infuse blood with cell-specific antibodies capable of killing the leukemic cells. There has likewise been recent progress in the use of epidermal growth factor (EGF) receptor blockers, which we will discuss in more detail in the following Development of Medications chapter.[42] Biological therapy such as interferon has also proved therapeutic.[43]

Advances in the field of leukemia research suggests that doctors will be able to optimize treatments based on patients' gene profiles. In a study that monitored 6,817 genes in leukemia patients, researchers found 1,100 genes that appeared to distinguish the two leukemias. One set of fifty genes differentiated them with one hundred percent accuracy.[44] These might have been hereditary mutations or damages acquired through the environment. This differentiation is called a "class prediction." A class prediction suggests the clinical expression of the genes, for instance, how the patient will respond to drugs and how long they will survive. The scientists concluded:

> The results demonstrate the feasibility of cancer classification based solely on gene expression monitoring and suggest a general strategy for discovering and predicting cancer classes for other types of cancer, independent of previous biological knowledge (other risk factors).[45]

And it can, given the state of today's technology, suggest a treatment. Gleevec, also known as "signal-transduction inhibitor-571 (STI571)," is a fruit of the drug discovery revolution attributable to genetics and molecular biology. Gleevec selectively targets the genetic defects that cause chronic myeloid leukemia (CML) by blocking an enzyme called tyrosine kinase. It has been tested, approved, found effective and does not cause the side effects usually associated with chemotherapy.[46]

Dr. Brian J. Druker of the Oregon Health Sciences University in Portland studied Gleevec, and found it brought the white blood cell count back to normal within four weeks in fifty-three of the fifty-four volunteers with CML. "This is what cancer research has been waiting for," Druker said. "It is the beginning of a whole new era in cancer therapeutics." He continued, "If we understand the critical abnormalities that drive a cancer, we can target the cancer with an effective and non-toxic therapy. We need to identify the critical abnormalities in each and every cancer so drugs like STI571 can be developed for each cancer."

Dr. Lou Fehrenbacher, a hematology/oncology specialist at Kaiser Permanente Medical Center in Vallejo, Calif., said in an e-mail to ABCNEWS: "The major importance of this drug is the nature of its discovery [it is a logically engineered creation]. . . . The enzyme was isolated, reproduced, and then a drug to inhibit it was engineered." Paul A. Bunn Jr., president-elect of the American Society of Clinical Oncology, said, "Read my lips, this is real, not mice."[47]

Gleevec is classic *in vitro* research, an example of what scientists can do when they use the proper research methods. Druker did test his discovery on animals after years of studying test tubes and cell cultures, but whether Gleevec cures leukemia in animals or not is irrelevant, because it is made to order for humans. We are sure it will work in some animals and not be effective in others. Drugs like this sometimes skip animal tests altogether and go directly to clinical trials. If they are tested on animals, they go to clinical trials regardless of the outcome in animals.

Animal models are not going to cure leukemia. Dr. Newlands wrote in *Successes, Failures and Hopes in Cancer Chemotherapy*:

> Finding a particular agent that is effective against an animal leukaemia provides no predictive value that it will be effective against either a human leukaemia or a different human tumour. . . . Unfortunately, none of the [animal] models we have at the moment gives us quite the answer we need.[48]

Instead of wasting taxpayer and charity money on animal models, why not fund research that works like the methodology that resulted in Gleevec.

Cancer

One in every three women will have a diagnosis of cancer in her lifetime, and the figure for men is one in two. That makes cancer one of this era's greatest medical challenges. There was a fifty percent rise in cancer cases in the United Kingdom since 1971.[49] And according to the US Office of National Statistics, cancer here is rising faster than can be accounted for by the aging of the population. Although combating cancer has been a national priority for thirty years, the public is still startlingly unaware of its causes and how to prevent it. In part, the cancer research community's predilection for the animal model is responsible for that naiveté.

Again and again, by contrast, human-based investigation uncovers real, applicable information.

Notably, cancer is not a single disease, but a way of classifying many diseases. Yet, all cancers have one aspect in common. That is the untrammeled proliferation of abnormal cells, cells that can invade and destroy healthy tissues. All are of one of three types—*sarcoma, carcinoma,* or *leukemia/lymphomas.* Sarcoma occurs in connective and supportive tissues. Carcinoma attacks the skin and the lining of body cavities and organs and in the glandular tissue of the breast and prostate. Leukemia and lymphoma, which we have already discussed, overtake blood-forming tissues and invade lymph nodes as well as bone marrow. All told, there are at least two hundred cancers in humans alone, and animal versions are, though in some cases similar, not the same diseases.

The critical message that cancer research expenditures have failed to convey about these diseases is that people have far more control over their vulnerability to cancer than one might believe. Most of us can keep our risk of cancer to a minimum by living healthfully. Mistakenly, many people believe that frequent exams will decrease the likelihood of cancer. They will not. Only good health practice will. You can catch cancer earlier by having regular checkups and maybe improve your chances of survival or you can catch a precancerous lesion before it turns malignant, but regular checkups do not prevent cancer *per se.* There are, however, better ways to live that prevent cancer. We can minimize exposure to carcinogens. We can diminish our intake of substances that deplete natural immunity. We can exercise. We can eat well and assure that our diets build up our immune system. These activities reduce our chances of getting cancer. Exams only spot cancer, the hope is, when it is still treatable. Again, exams are worthwhile. But they do not *prevent* cancer.

Blind to the idiosyncrasies that cause cancer to manifest we tend to think cancer is an irrational manifestation of disease. In the past, pathologists classified tumors according to their morphology—that is, their form and structure. The limitations of this kind of classification are that it does not reliably convey any given tumor's clinical course, the patient's life expectancy, or the tumor's response to a given treatment. It left oncologists with only a generalized cancer treatment wherein they indiscriminately blasted the tumor with radiation and/or subjected the patient to chemotherapy. This shotgun approach is a treatment of the past. New technology affords a far more differentiated picture of cancers that were hitherto indistinguishable.

We know that cancer is an uncontrolled rate of cell growth. Cancer can result from mutations that stimulate expression of the oncogenes or from mutations that inhibit expression of the tumor-suppressor genes. An oncogene is a gene that stimulates cell growth. It is like the accelerator on a car. If the oncogene is mutated the accelerator is stuck. Tumor-suppressor genes are the brakes. The gene *P53* is a tumor-suppressor gene. In roughly fifty percent of human cancers, it is inactivated, so, no brakes inhibit cell growth.[50] As an example, patients with ovarian cancer sometimes have a mutation in the *P53* tumor suppressor gene. Some were treated with chemotherapy and gene therapy with good results. The patients lived longer. (The researchers injected the gene with an adenoviral vector, as described earlier in this chapter. Clinical trials like these are producing results.)

Other cancers that have been linked to tumor-suppressor genes include neuroblastomas, retinoblastoma, pheochromocytoma, and Wilms' tumor. Many human cancers occur in humans with abnormalities at chromosome *11Q2224*. This suggests that this area of chromosome 11 contains a tumor suppressor gene, which if inactivated may lead to cancer. One gene in that region, *PPP2R1B*, has been found that is associated with certain cancers of the colon and lung.[51]

Heredity, which can be tracked epidemiologically and through *in vitro* research, has a role in an estimated twenty percent of all cancers. Some forms of breast cancer, as an example, runs in families. European families that have a mutation in *CDKN2A*, a multiple tumor suppressor gene, were found to be at increased risk of melanoma, pancreatic cancer, and breast cancer.[52] Other familial pancreatic cancer points to another specific mutation that predisposes them to the disease.[53] Colorectal cancer is common in families whose genes predispose them to polyps in the colon. In another example, a mutation in the *SRD5A2* gene appears to contribute to the development of prostate cancer in African American and Hispanic men.[54] (A missense substitution is a mutation that renders the gene sequence nonsensical.) Sometimes, as when DNA has a high frequency of breakage, cancer is almost inevitable. In other cases, the risk exists, but can be averted by a healthful lifestyle.

Sometimes heredity is not to blame. Exposures can create genetic damage, also leaving us vulnerable. As an aspect of the Human Genome Project, the National Cancer Institute has developed a program called the Cancer Genome Anatomy Project (CGAP). The CGAP is compiling a comprehensive record of genes involved in human cancer as they appear. Using a technology called the DNA microarray, scientists compare the activity and interaction of genes. The DNA microarray, also known as a gene chip, is a glass slide with tiny wells coated with fragments of DNA from different genes.[55]

When genes are "active" they make a chemical called messenger RNA (mRNA) that tells the cell to make a particular protein. So, to see if the genes are active, scientists test for the presence of mRNA. Gwen Z. Acton, Ph.D., who is assistant director of the Functional Genomics Program at the Whitehead Institute/MIT Center for Genome Research in Cambridge, Massachusetts, explained the use of DNA chips to appraise gene activity:

The amount of RNA produced by the gene generally corresponds to the amount of product produced by the gene. It is therefore possible to estimate the amount of gene expression by measuring the amount of RNA produced. The ability to measure gene expression has been facilitated in recent years by the availability of DNA microarrays or DNA chips, which consist of a matrix of DNA spots on a glass slide, each one corresponding to a single gene. Using DNA microarrays, it is possible to monitor expression of tens of thousands of genes simultaneously, throwing open the possibility of unprecedented molecular characterization.

Gene expression studies can detect not only inherited variation in gene expression, but also responses to environmental influences or stochastic changes in the patient. They are therefore potentially good tools for: (1) identifying additional potential drug target candidates; (2) providing more accurate classifications of disease; (3) predicting clinical outcome and presymptomatic disease; and (4) monitoring and predicting drug responses. To identify new drug targets, DNA microarrays are used to compare the expression of genes in two different cell, tissue, or tumor types. The genes that are more highly expressed in one cell type, but not in the other, are candidates for key determinants of the difference. For example, in one study, metastatic vs. non-metastatic cells were compared using DNA microarrays. Some genes, such as *RhoC*, were preferentially expressed in the metastatic, but not in the non-metastatic cells. It was subsequently shown that *RhoC* is necessary and sufficient to cause metastasis, leading to a better understanding of the process by which cells become metastatic. This type of analysis can lead to improved drug development, since genes that are preferentially expressed in disease tissue are typically good candidates for drug targets.[56]

Hence, DNA microarray testing helps identify subsets of tumors with distinctive molecular profiles that are often imperceptible to the naked eye. These distinctions are important since they describe different causes of cancer, and those cancers demand different therapies. They are also prognostic, giving oncologists information on the progression of the disease.[57,58]

Scientists at NIH, using *in vitro* techniques and cancerous tissue obtained from patients, have developed a new genetic test that is able to distinguish between hereditary and sporadic forms of breast cancer. This knowledge will be used to determine treatment. Researchers use a new technique called *gene-expression profiling* to differentiate between the two kinds of breast tumor. The test also uses DNA microarrays. It allows the researchers to assess the activity of six thousand genes within breast cancer cells. The test makes approximately 250,000 measurements.

A serial analysis of gene expression, called SAGE technology, evaluates gene expressions in different tissues. Researchers identified a set of forty genes that were expressed at elevated levels in cancer tissue, but are not seen in normal tissue. This set may provide diagnostic markers and targets for therapy.[59] Physicians can sometimes detect cancer through the presence of a specific enzyme, rather than subjecting the patient to a genetic analysis. For instance, ovarian cancer has been often diagnosed late because it is usually without symptoms until it is advanced. A new enzyme-based assay appears to detect ovarian cancer with twenty-five percent more accuracy than the standard test, the CA 125. The enzyme is lysophosphatidic acid.[60] Another protein called soluble Fas has been found circulating in human blood, and it too may provide a better marker for earlier predictions of ovarian cancer.[61]

In the majority, however, genetic abnormality, whether hereditary or exposure-related, is not enough, in and of itself, to create illness. Again, it is how we treat our bodies that either fortifies them against cancer or abuses them into cancer. Living in a way that effectively censors the onset of cancer is no longer considered "alternative medicine." The world's top authorities agree. The president of the American Cancer Society, Dileep G. Bal, MD, stated:

> Two-thirds of cancer deaths in the US can be linked to tobacco use, poor diet, obesity, and lack of exercise . . . *poor diet is quantitatively an equivalent cancer risk factor to tobacco* . . . high fruit and vegetable consumption reduces the risk for at least ten different cancers, including cancers of the lung, stomach, colon, esophagus, and larynx. Also, evidence is mounting that increased consumption of legumes and grains reduces risk for stomach and pancreatic cancers. Red meat consumption is implicated in the development of certain cancers . . . total fat intake and saturated animal fats are linked to the occurrence of hormone-related, lung and colorectal cancers [emphasis added].[62]

So, two-thirds of all cancers are entirely avoidable, not because doctors can protect us, but because we can protect ourselves. Vegetarians have about half the cancer risk of meat-eaters for good reasons. A vegan diet (no animal products, that is, no meat, eggs or dairy) increases the amount of sex-hormone binding globulin (SHBG) in the blood. And SHBG protects against hormone-related cancers.[63] Likewise, human studies suggested that high levels of the insulin-like growth factor IGF-I play a key role in the pathogenesis of prostate cancer, and men who eat a vegan diet have lower levels of IGF-I (the insulin growth factor). High-fat foods and alcohol effectively nudge cancer forward. Like tobacco smoke and certain chemicals, these negative dietary influences are called "cancer promoters."

Human-based research has made nutrition's role in cancer an irrefutable fact, while animal-based nutrition research typifies the ineffectiveness of animal-model protocols. For instance, the need for vitamins A, D and nicotinic acid—so powerful in maintaining immunity—was discovered in humans. Even when the human evidence is already conclusive, researchers who use the animal model try to ride the nutrition bandwagon . . . always with mediocre results. Studies found that the only animals that do not synthesize ascorbic acid (vitamin C) are primates (including humans), and guinea pigs.[64] Poor parity is not surprising since humans do not thrive on the same foods as rodents, or dogs, or even other primates. Not just the nature of the food, but also metabolism, amount, and feeding intervals are different. Dr. David Conning, director general of the British Nutrition Foundation admitted in 1991:

> Although the experimental animal is deficient as a model of human nutrition, it has been used extensively in studies to elucidate the effects of nutrients on experimental carcinogenesis.[65]

This is like saying, "the animal model is a waste of money but we use it anyway." Animal models in nutrition research persist.

Additionally, in those avoidable two-thirds of all cancer cases, exercise is another preventative that we did not require animal models to spell out for us. A new analysis of data from the Nurses' Health Study confirms that women between the ages of thirty and fifty-five years can reduce breast cancer risk twenty percent by exercising one hour each day.[66] Similar data analysis shows a sedentary lifestyle and obesity as contributing factors in other cancers. We do not need to put mice on treadmills to confirm this data. No single factor—exercise or diet or weight or substance avoidance—is in and of itself a magic bullet. But each factor improves chances for people who, through exposures or heredity or both, are susceptible.

There is agreement that some forms of cancer are preventable, but cancer's preventability gets short shrift while researchers vigilantly pursue cancer's causes, too often trying to obtain their clues from animal data. Continually, animal-modeled data are either at odds with known human data, or in the absence of previous human data, suggest an avenue that leads to a dead end. Both outcomes make research unnecessarily circuitous and jeopardize human lives. Human-based research, by contrast, provides real, predictive data.

Early cancer researchers had no sophisticated instrumentation and yet they did not rely on animals. They wanted to answer fundamental questions, such as, is cancer contagious? Over the years, bold self-experimenters demonstrated that cancer is not communicable. In 1777, Dr. James Nooth took cancer tissue from his patients and implanted it under his skin. In 1808, Dr. Jean Louis Alipert injected cancerous tissue into his body as well. In 1901, Dr. Nicholas Senn placed cancerous tissue beneath his skin. We have already mentioned Brittingham's self-experimentation with leukemia. Because these experiments did not in-volve animals, the medical establishment then refused to honor their conclusions. Nevertheless, we now accept their findings as accurate. These men's bravery has done more for our knowledge base than using animal models has ever done.[67]

The less information there is, the more people tend to generalize, as we examined with our discussion of genetic proximity with apes. Not all that long ago, physicians believed that viruses caused cancer because they learned that at least some bird cancers result from viral exposure. Then more research on animals proved that some animals did not re-spond to viruses as birds do. Again, different animal models gave very different results. Afterward, the virus theory was totally abandoned. Four decades later when a virus was found to cause cancer in humans, the medical community finally had conclusive evidence that some viruses could cause cancer in humans. As we have said, test enough different species and you will eventually find one to mimic humans. But how do you know in advance which one? Most of the viruses associated with animal tumors are retroviruses, but those that cause cancer in one species may not in others. For instance, the virus SV40 does not cause cancer in monkeys but does so in hamsters.[68] In humans, the Epstein-Barr virus, hepatitis, and papilloma viruses can all lead to cancer but do not result in cancer in most animals.

As we have already mentioned, radiation causes cancer. Though ani-mals too suffer from cancer after exposure to radiation, research that repeats this effect is superfluous because the cancers are not the same and because there is human evidence in plethora. We have Mme. Curie's demise and thousands of other examples in addition to a molecular explanation to back this up. The same concept holds true for chemical influences that can, even in a single exposure, produce irreversible changes in DNA. These changes—often exacerbated by bad diet, too much alcohol and other life-depleting habits—manifest in cancer. Epi-demiology, by contrast with animal indicators, has been a consistently great identifier of carcinogens.

The problem, as we have said, is that even after overwhelming human data exist, those with a vested interest in animal models still feel they need to recreate specific conditions in animals. Why? These efforts al-most always show a range of response that is anything but conclusive. That very lack of conclusiveness allows whoever is profiting from the carcinogenic enterprise to persist. For example, in 1776 Percival Pott determined that coal tar caused cancer because so many chimney sweeps got cancer. But coal tar did not give cancer to animals. Two centuries later, while coal miners continued to expire, Yamagiwa and Ichikawa managed to give cancer to some animals using coal tar, but even then rabbits did not respond as mice did.[69] Still in retrospect, the animal models misled. Human-based studies could have offered information that answered questions, not introduced new and irrelevant quandaries.

In their book *Higher Superstition*, Paul Gross and Norman Levitt write that the government now acknowledges that some substances proven carcinogenic in rats are not carcinogenic in humans. They wrote:

> The *Federal Register* for July 17, 1992, carries the extraordinary announcement that large numbers of substances, classified heretofore by the National Toxicology Program as carcinogens, are to be removed from that list. This announcement codifies understandings that have been growing slowly—and against bitter political opposition—in the scientific community, to the effect that rodent-to-human extrapolations used in animal screening programs are invalid.[70]

To put an end to costly and nonpredictive research, many scientists suggest that carcinogenicity testing be compared along a tenet called "weighted values." Weighted correlations assign larger value to tests that predict the carcinogens that kill millions like smoking and less value to those tests that predict fewer mortalities. Take, for instance, the epidemiological evidence that indicates the high percentage of all smokers who get lung cancer. By contrast, tobacco-exposed rats do not get lung cancer. Which is the more reliable research vehicle? Epidemiology, clearly, not animal experimentation. William Campbell, president and CEO of Phillip Morris testified under oath in 1993:

> Q. Does cigarette smoking cause cancer?
> A. To my knowledge, it's not been proven that cigarette smoking causes cancer.
> Q. What do you base that on?
> A. I base that on the fact traditionally, there is, you know, in scientific terms, there are hurdles related to causation, and at this time there is no evidence that they have been able to reproduce cancer in animals from cigarette smoking.[71]

Campbell said, effectively, that cigarette smoking does not cause cancer in animals. We all know it causes cancer in humans though. If weighted value were actually put into effect, animal models for carcinogenicity would end because they are such poor predictors. Scientists wrote the following:

> For decades the clinical observation of an association between cigarette smoking and bronchial carcinoma was subject to unfound doubt, suspicion, and outright opposition, largely because the dis-

ease had no counterpart in mice. There seemed no end of statisticians craving for more documentation, all resulting in the fateful delay of needed legislative initiative.[72]

They admitted that exhaustive attempts to fill mouse lungs with malignancies have been unrewarding. In fact, tests done on rats and mice agree with each other only seventy percent of the time, and agree with humans far less. It is now a well-known fact that smoking does not cause cancer in animals unless very exaggerated conditions are imposed and even then it is difficult to make a mouse get lung cancer from cigarette smoke.[73] Dr. Michael Utidjian of the Central Medical Department, American Cyanamid Company in New Jersey, writing in the book *Perspectives in Basic and Applied Toxicology* stated,

> Surely, not even the most zealous toxicologist would deny that epidemiology, and epidemiology alone, has indicted and incriminated the cigarette as a potent carcinogenic agent, or would claim that experimental animal toxicology could ever have done the job with the same definition.[74]

Since lab rodents do not get cancer from the same exposures as humans do, even in vastly greater dosages, scientists crafted means of giving it to them. The mouse destined to become a "xenograft mouse" has human cancer inserted into it. Afterward, scientists watch what happens. By contrast, the recipe for a type of transgenic animal, the "oncomouse," calls for inserting the genes from human cancer victims into embryonic mice who then reproduce the genes. This leaves scientists with an immense reservoir of rodents on which to test their theories, which would be great if it worked. However, as Dr. Tyler Jacks of the Massachusetts Institute of Technology states:

> One might expect that these animals would mimic human symptoms, not just the genetic mutations. In fact, that is usually the exception, not the rule . . . the genetic wiring for growth control [cancer growth] in mice and humans is subtly different . . . Animals apparently do not handle the drugs in exactly the same way the human body does.[75]

Nevertheless, not since the "farmer's wife with her carving knife" has there been so much commotion over mice. In 1999, the NCI sought out nineteen new groups of mouse geneticists from more than thirty American institutions for the Mouse Models of Human Cancers Consortium, headed by Dr. Tyler Jacks himself. Of it, NCI director, Dr. Richard Klausner said:

The consortium will provide a new interactive research platform to test the validity of these models to elucidate the development and behavior of cancer, to test new approaches to detection, diagnosis, and imaging, and to evaluate prevention and treatment.

Is this the same Richard Klausner who said to the *Los Angeles Times* the year before, in 1998:

The history of cancer research has been a history of curing cancer in the mouse. We have cured mice of cancer for decades, and it simply didn't work in humans. [76]

As cancer researchers carry their animal investigations from the live species into the test tube, they find incongruities between animal and human data:

Primary rodent cells are efficiently converted into tumorigenic cells by the coexpression of cooperating oncogenes. However, similar experiments with human cells have consistently failed to yield tumorigenic transformations, indicating a fundamental difference in the biology of human and rodent cells.[77]

In studies conducted by the National Cancer Institute, drugs effective against human cancers did not work against those same cancers in mice 63 percent of the time. Even *Lab Animal* magazine stated:

Mice are actually poor models of the onset of the majority of human cancers despite the reliance of biomedical research on mouse models to understand these cancers. Genetically altered mouse models of cancer carry the cancer-causing mutation(s) in every cell. But apart from hereditary cancers, the majority of human cancers are the result of sporadic mutation in oncogenes and tumor-suppressor genes. . . . [78]

Monkeys, considered more human like, also frustrate cancer researchers. According to one pro-animal model textbook:

Spontaneous tumors in monkeys are very rare. . . . Many researchers believe that monkeys have an inherent specific resistance to malignant tumors. The low incidence of spontaneous tumors in monkeys has been associated with difficulties in experimental induction of tumors in these animals.[79]

This same source, by a researcher who uses animal models Dr. Dzhemali Beniashvili, in numerous places describes the limitations of his protocol in studies of tumors throughout the body.

Certainly, we have tens of thousands of humans who suffer from cancer. Tissues from these patients yield far more accurate results than those from compromised animals. Yet, we continue to fund experimentation on animal models when they do not give way to useful data. This is not helping us find a cure for cancer. As oncologist Dr. Harrison writes:

> It is in fact hard to find a single, common solid neoplasm [cancer] where management and expectation of cure has been markedly affected by animal [modeled] research. Most human cancers differ from the artificially produced animal model . . . [80]

Even a very pro-animal model article in the *Journal of the National Cancer Institute* stated:

> . . . [animal models] may not offer an uncomplicated straightforward means of discovering preventable causes for the majority of human cancers, and at the very least it certainly does not seem likely that they can offer a reliable means of estimating quantitative human hazards.[81]

On the other hand, again, human observation offers *consequential* data. For instance, we now know that some people have an enzyme that detoxifies tobacco smoke; this protects them from lung cancer. A genetic variant in some people renders this enzyme less effective, which research determined can lead to lung cancer from secondhand smoke.[82] The errors work the other way too. Fifteen years after inducing cancer in normal mouse cells, scientists decided to attempt the same thing in human cells. And it didn't work.[83]

Carcinogenicity studies designed by researchers who use animal models are so contrived that it is impossible to draw meaningful conclusions from them. For example, we know that disrupted endocrine activity can lead to cancer. Experiments on mice showed huge doses of a chemical found in plastics, bisphenol-A, throw murine endocrine systems off kilter. The implication was that pregnant women should avoid foods packed in bisphenol-A plastic because residual amounts of the substance might cross the placenta and affect highly sensitive embryos. Attempts to find human relevance were foiled though. In order to receive the equivalent dose of bisphenol-A, a 132-pound woman would have to eat 1,265 pounds of plastic wrapped food per day. The journal stated the obvious, that "the relevance of these findings to humans is moot—after all, mice and humans are not the same."[84,85] Certainly, this is not a good use of research revenues.

Dr. Lester Lave, of the Carnegie-Mellon University in Pittsburgh, and colleagues Drs. Ennever, Rosenkrantz, and Omenn, writing in the journal *Nature,* characterized the problem:

Extrapolating from one species to another is fraught with uncertainty. . . . For almost all of the chemicals tested to date, rodent bioassays have not been cost-effective. They give limited and uncertain information on carcinogenicity, generally give no indication of mechanism of action, and require years to complete. [They are] rarely the best approach for deciding whether to classify a chemical as a human carcinogen.[86]

And Philip H. Abelson who, as Deputy Editor of *Science* magazine receives a lot of data across his desk, stated:

The principal method of determining potential carcinogenicity of substances is based on studies of daily administration of huge doses of chemicals to inbred rodents for a lifetime. Then, by questionable models, which include large safety factors, the results are extrapolated to effects of miniscule doses in humans. . . . The rodent MTD test that labels plant chemicals as cancer-causing in humans is misleading. The test is likewise of limited value for synthetic chemicals. The standard carcinogen tests that use rodents are an obsolescent relic of the ignorance of past decades.[87]

Larger mammals provide no better indicators. Dr. C. Parkinson of the Centre for Medicines Research, Surrey, England, and Dr. P. Grasso of the Robens Institute of Health and Safety, Surrey, England stated in 1993:

Why the dog was ever considered as an appropriate animal for carcinogenicity testing is also not entirely clear. . . . Despite the obvious problems of study design and interpretation, carcinogenicity tests in the dog, lasting 7 years, were requested by regulatory authorities from the late 1960s. . . . One of the best known examples of the inappropriate use of the dog was the carcinogenicity testing of hormonal contraceptives. It is now understood that mammogenesis in the dog is very different from that in primates; quantitative and qualitative differences exist in the feedback control mechanisms, receptor content and behaviour, and target sensitivity and responsivity. As a result of this biological difference there was a high incidence of mammary tumours in long-term studies in dogs treated with progestagens/contraceptive steroids such as lynestrol. Ultimately pressure from the scientific community led, relatively recently, to the requirement for carcinogenicity studies in dogs being dropped.[88]

Next to lung cancer, the second-highest mortality rate among cancers is colorectal cancer. An estimated 156,000 Americans will be diagnosed with colorectal cancer this year, and 57,000 will die. The enormity of this threat creates demand for studies of colorectal cancer. Looking at research in this area is exciting, because it underscores the role each one of us plays in keeping our bodies clear of cancer. It stands to reason that diet would strongly influence tumor growth in the colorectal region in particular, and epidemiology has many times illustrated that certain foods affect the colon positively and negatively. As an example, fiber is known to decrease the risk of colorectal cancer. But such is not the case in animal models. Other factors contribute to whether the animal contracts cancer, as the following indicates:

> The expected number of tumors even in the control group can be manipulated by altering the amount and timing of the dose (for example, before or after test food); alternating the number of doses, and how they are administered (e.g., mouth, rectum injection); and changing what will be fed in addition to the test fiber or food. Many experiments show that things not considered important made a crucial difference in the result—for example, caloric sufficiency. Should the animals eat *ad libitum* or should the calories be restricted? The possible number of experimental designs would be in the thousands.[89]

Meat and dairy products compromise the colon, as we have said, and colorectal cancer is three times more common in meat eaters.[90] A recent study of South Africans found that colon cancer is seventeen times more prevalent in whites than in blacks. The reason for this, the study concluded, was the absence of animal products from the blacks' diet.[91] Despite the growing evidence, less than one-third of all colorectal cancer patients reported that their physicians suggested diet changes.[92] This seems an indefensible oversight when one considers the extent of data supporting the links.

More than ninety percent of all colorectal cancers can be cured if they are caught in the earliest stages, before symptoms manifest. So, exams, though not preventative, are critical. Colonoscopy is the standard diagnostic screening procedure. DNA-based screening may detect almost all colorectal cancers. Noninvasive tests, such as the k-ras screening, indicate the presence of a colorectal tumor by identifying mutated genes, which are shed from the lining of colorectal polyps and tumors in fecal samples.[93,94]

The majority of HNPCC (hereditary nonpolyposis colorectal cancers) are caused by mutations in genes *MLH1* and *MSH2*. (These mutations also signal an increased risk of endometrial cancer.) If the mutations are found, the patient can undergo colonoscopy and/or hysterectomy much earlier than otherwise recommended. We also know that some colorectal cancer patients have another sort of genetic flaw—an additional functioning copy of the gene that controls the production of a substance called insulin-like growth factor-2 (IGF2). This mutation causes colorectal cells to grow too fast. Scientists do not yet know whether the genetic defect causes the cancer or even whether it is there before people manifest symptoms.[95] And, as we have said, a great number of people's genes predispose them to grow polyps in the colon, and these can lead to cancer. So those genes too are red flags. Having come this far, some scientists still detour through animal testing that, very likely, will produce unusable results.

When colon cancers are forced upon rodents by induction, they are deviant, behaving very differently than they do in humans. One difference is this: Rat-colon tumors kill by obstructing the colon. Human-colon cancer kills by metastasizing to other places in the body. Tumors of the rat bowel do not usually spread. In rats, it is most often the small bowel that is affected; in humans it is the large bowel or colon. Despite these dramatic differences between species, we call both diseases "colon cancer" and proceed with millions of dollars-worth of tests on rodents to obtain information, information that cannot be extrapolated to humans. Is a cancer the same cancer, as long as it is in the same location? No. Though researchers, institutions and rodent vendors stay profitable, the animal model takes money away from research and innovation that will hasten the development of non-invasive screenings. Dr Jeffrey E. Green of the National Cancer Institute's Laboratory of Cell Regulation and Carcinogenesis stated that:

> . . . some findings in colon cancer mice, which were very good models, actually led to clinical trials in humans which resulted in an increase in cancer.[96]

If a model that leads to drugs being released that increases cancer could be good, we hate to contemplate what might need to occur for a model to be considered bad.

Colorectal tumors have a high probability of recurrence. Scientists have found, through human studies, that this probability is at its highest in patients with above average levels of a certain carbohydrate antigen in their carcinoma.[97] Knowing this, physicians can put these patients on special regimens and conduct more frequent colonoscopies. Unchecked, colorectal cancer frequently metastasizes to the liver, measurably increasing the possibility of fatality. Administering chemotherapy directly to the liver in these patients improved the outcome in a human study—another example of actual human-based progress.[98]

Diagnostics are improving for other forms of cancer too. In the past, the best test for cancer a physician could conduct would be to examine a biopsy of the lymph node nearest the growth for cancerous cells. This tedious and not terribly accurate test often overlooked early stage cancers. Now using the reverse transcriptase-polymerase chain reaction (RT-PCR) assay for tyrosine messenger RNA, oncologists can be far surer of their diagnosis. They use a PCR (polymerase chain reaction) to duplicate a single strand of DNA billions of times in a few hours *in vitro*.[99] This allows them to detect the cancer even when it is very small.

Just as human epidemiology leads us to cancer's causes and *in vitro* technology to better diagnostics, clinical observation and other human-based investigation direct us to cancer's treatments. "Chemotherapy" began in 1946 when Alfred Gilman and Frederick Philips convinced the world that nitrogen mustard, derived from the mustard gas of World War I could cause lymphomas to shrink. In 1949 Sidney Farber and his colleagues showed that methotrexate could induce remissions in children with leukemia. These early indications suggested chemotherapeutic avenues, but using animals to hone them frequently frustrated them. In 1965, Dr A. J. Shorthouse and colleagues wrote:

> Most available chemotherapeutic agents have been developed using serially transplantable rodent tumours. Unfortunately their biological behavior and chemotherapeutic sensitivities do not closely resemble those found in human solid tumors. Chemotherapeutic data derived from these experimental systems may therefore be misleading with the result that patients in clinical trials frequently receive ineffective agents.[100]

Dr. Irwin Bross later corroborated this view.

> ... the discovery of chemotherapeutic agents for the treatment of human cancer is widely heralded as a triumph due to the use of animal model systems.... There is little, if any, factual evidence that would support these claims.... Indeed, while conflicting animal results have often delayed and hampered advances in the war on cancer, they have never produced a single substantial advance either in the prevention or treatment of human cancer.[101]

Chemotherapy has indeed been a modern-day miracle for some cancer victims. But for others it has been as bad as the disease itself. Chemotherapy is toxic and often depletes the body's overall health as it attacks the cancer. Also it is not always successful. Chemotherapy treatments for testicular cancer have always been particularly harmful. Therefore, cancer research continues to look for more effective, less damaging treatments. The idea to combine treatments was based on clinical observation.[102,103] Happily, Lance Armstrong, who won the Tour de France in

1999, 2000, and 2001, participated in trials on a less toxic combination of drugs and was victorious over his cancer.

As the years go by, chemotherapies become more sophisticated. Direct drug delivery is sometimes an option for many types of cancer, if it is detected in early stages. Retinoids, which are synthetic variations on vitamin A, can be effective in arresting tumor growth. Researchers are experimenting with new aerosol agents to deliver concentrated retinoids to the lungs.[104]

Radiation, the other workhorse of cancer treatment, was also discovered as a consequence of human observation. Over 37,000 American men die from prostate cancer every year, making it the leading cancer killer of men after lung cancer. Many prostate cancer patients are treated with radiation therapy. Without performing further invasive surgery, physicians can determine if the radiation was effective by intermittently gauging levels of prostate-specific antigen (PSA).[105] This knowledge, thanks to *in vitro* research, increases patients' confidence that they are no longer at risk.

Epidemiology and *in vitro* research recently identified the first known gene putting men at high risk for developing prostate cancer.[106] Scientists discovered that two mutations of the gene named Hereditary Prostate Cancer 2 (*HPC2*) increase the risk. To find the gene, the scientists studied the Utah state cancer registry and the Utah Population Database. They isolated families with a history of increased rates of prostate cancer. They obtained blood samples and analyzed DNA. Thus, they found *HPC2*. Lisa Cannon-Albright, a genetic epidemiologist at the University of Utah and one of the scientists involved in the study supported the gene-based approach, wrote:

> I believe all common diseases have an underlying, inherited predisposition. Because there are other risk factors and multiple genes for common diseases, these types of genes are thought to be extremely difficult to find. But finding them is clearly possible, and is going to provide the key to earlier diagnosis and more appropriate treatment. That is going to result in less illness and death.

In one experimental development, scientists harvested prostate tumor cells and stimulated their white blood cell component to grow in a culture. They then irradiated the cells to ensure that they would not continue to grow. When they injected this transfected tumor cell "vaccine" into human patients, they found that it activated the entire immune system. Scientists concluded that these and other data "seem to suggest that systemic immune responses to human prostate carcinoma can be generated against a tumor type that has been conventionally viewed as being nonimmunogenic, and refractory to immunotherapy."[107] A prostate cancer vaccine would vastly reduce the incidence of cancer among men.

Oncologists continue to search for ways to get around the limitations of chemotherapy and radiation. Malignant cells need oxygen and nutrients, just as healthy cells do, and they require a blood supply to deliver these. Some tumors make their own blood supply in a process called angiogenesis. Hence, much cancer research pivots around agents that may inhibit angiogenesis.

There is proof that many chemicals, including endostatin, angiostatin and thrombospondin, inhibit the growth of blood vessels needed for newly forming cancers in mice. Scientists get very excited when there is progress in this venue; however, efforts to reproduce these results in humans have thus far failed. Now, human-based evidence indicates that the reason for this may be that human tumors differ. In humans, tumors grow their own blood supply, while the mouse tumors stimulate the blood supply to nurture them. One of the discoverers, Andy Maniotis of the University of Iowa, pointed out, "People are very complacent with their animal models. But this [evidence] begs the question of whether there exists a good model of cancer."[108,109]

One lab species, the "nude mouse," has been used for years to model human tumors. The nude mouse has certain deficiencies of the immune system that allow scientists to transplant human tumors into it, where they then grow. The nude mouse effectively becomes a sort of live Petri dish for human tumors. This paradigm presumes a lot. Oystein Fodstad of Norwegian Radium Hospital in Oslo wrote in *The Nude Mouse in Oncology Research*:

> The relevance of studies in the biology of human tumors, when grown as xenografts in nude mice, is based on the assumption that the interaction between the tumor cells and the host environment is closely similar to that in patients. One striking difference between xenografts and their parent tumors is, however, that the blood vessels and stromal cells in the xenograft are of murine origin. It would, therefore, not be surprising if this had significant impact on important characteristics of the human cells in the tumor. . . . The lack of functional T cells permits the growth of most human tumors, but it has been shown that T cell-like activity may be induced that probably affects tumor take. Whereas the B cell function in nude mice is relatively unaffected, natural killer (NK) cell activity is high. Moreover, the activity of the macrophages, particularly in the peritoneal cavity and in the lungs, may be different from that in man. It is conceivable that such differences may interfere with different types of biological studies in nude mice. This is clearly possible in chemotherapy experiments with drugs that affect the immune system.[110]

Fodstad goes on to explain why these mice models fail:

> The apparent lack of spontaneous metastasis formation from s.c. [sub-cutaneous] tumor xenografts in nude mice, represents a major

limitation for studying this important aspect of tumor biology. . . . Xenografts in nude mice provide the best available model for studies on human tumor biology *in vivo*. The usefulness of the model is, however, limited by several factors inherent to either the host or the tumor, or the interaction between the two. There are many differences between the tumors that have been transplanted into nude mice and the ones still in human. For example: cell cycle parameters, growth rate, metastatic spread, invasive properties, origin of stroma, pharmacokinetics, and metabolism of the tumor-bearing host are different from the situation in the cancer patient.[111]

Dr. Fodstad is certainly not alone in his doubt. The National Cancer Institute harbored mice that were growing forty-eight different kinds of human cancers. Scientists treated them with twelve anticancer drugs that were currently and effectively used in humans. In thirty out of the forty-eight, the drugs did not work. That means that in sixty-three percent of the time the mouse model was not predictive. Oncomice, who had the gene causing the cancer inserted, did not predict human drug response much better. Mice who have the genes for colon cancer, retinoblastoma, and other cancers just do not replicate the human condition. Of xenograft mice and oncomice experiments Edward Sausville, associate director of the division of cancer treatment and diagnosis for the developmental therapeutics program at the National Cancer Institute, was quoted in *Science* as saying, "We had basically discovered compounds that were good mouse drugs rather than good human drugs."[112]

Obviously, we cannot have researchers inserting human tumors into humans. But we do not need to; there are plenty of human cancer victims to observe. Human evidence from the thalidomide tragedy in the 1950s proved that thalidomide inhibits blood vessel formation as well as causing terrible birth defects. A multiple myeloma patient received successful treatment with thalidomide in 1965. Subsequent *in vitro* and clinical data led physicians to try the drug on multiple myeloma patients again, but not until 1997 because thalidomide had been labeled a "dangerous drug."[113] It looks promising.

Each kind of cancer seems to demand its own treatment. Barnett Rosenberg watched bacterial growth diminish *in vitro* when electric current was passed between platinum electrodes. This paved the way for *cis*-platinum therapy for epithelial cancers.[114] Two human proteins—METH-1 and METH-2—also curtail the growth of endothelial cells. These are much more active than endostatin and thrombospondin in preventing blood vessel formation in humans. The two proteins were found in a systematic search of the human gene database. Though similar to one another, METH-1 and METH-2 are made by different genes. It is hoped that these proteins will be instrumental in new anticancer drugs.[115] Rubitecan is another example of nonanimal model research leading to a therapy particular to a specific cancer. Rubitecan is one of a new class of drugs known as camptothecins that repress cell division in pancreatic cancer cells by interrupting an enzyme. *In vitro* research scientists found these in a Chinese plant called *Camptotheca acuminata*.[116]

Counterpoints to these success stories are the fiercely adverse reactions to matrix metalloproteinase inhibitors (MMPIs). Problems with these anticancer (and osteoarthritis) drugs developed on animals, forced the manufacturer, Bayer, to cease the studies. Dr. Frank Armstrong, head of worldwide drug development for Bayer, called the action of the drug disappointing for the company as well as patients, whose cancer in many cases had greater progression than that in placebo groups. Bayer's Director of Oncology Research, Dr. Mel Sorensen said, "The finding was very surprising to us and very contrary to preclinical [animal] data which confirmed the drug inhibited tumors in rodents."[117] Though this is only one example, "disappointments" that result from animal experimentation are not unusual.

So, on the one hand, we have millions of taxpayer and charity dollars spent on animal models, with too-often fruitless or harmful results. On the other, genetic medicine combined with clinical observation, epidemiology, advances in technology and *in vitro* research, speeds the race for more worthwhile cancer diagnostics and treatments. Once scientists have identified which genes are at work in which cancers—using DNA microarrays, and subsequently, computer models—the distinctions accurately predict degrees of responsiveness to therapies and life expectancies.[118,119]

Human-based research also strengthens customization of therapy. Take, for instance, breast cancer. Some breast cancers—25 to 30 percent—result as an over-expression of a gene called *HER2*. These respond to the drug Herceptin (like the leukemia drug Gleevec), one of the first pharmacogenomic or gene-customized drugs to be approved. Herceptin is a monoclonal antibody designed to act on a specific destination. Hence, instead of giving large doses of poisons and killing all cells, the drug targets only the cells that need to be eradicated. (Read more about Herceptin in the Development of Medications chapter.)

Other breast cancers express for the estrogen receptor gene. Those respond to tamoxifen. For example, women with breast cancer may have a genetic expression of *BCL-2*. This gene indicates a favorable response to tamoxifen. On the other hand, patients with mutant p53 genes have a more lethal form of the disease and should be treated with adjuvant chemotherapy.[120]

Medical therapy exists for all types of cancer. In some instances, the progress has been truly extraordinary. Childhood leukemia and advanced testicular cancers were once uniformly fatal; today, most of these patients are cured. For other cancers, progress remains minimal and has moved little beyond surgical excision. For "liquid" tumors—the leuke-mias and lymphomas—and for certain unusual cancers such as germ cell tumors or childhood sarcomas, the principal treatment is multiagent chemotherapy. For some patients, this treatment will be curative. For a substantial portion of patients, however, these treatments will only be palliative. Patients who relapse after initial therapy have a much worse prognosis. For "solid" tumors—such as breast, colon, lung, and prostate cancer—multimodality therapy is the "gold standard." Treatment to extirpate these cancers at their source—usually surgical removal of the tumor—remains an essential step. Notably, refinements in surgical tech-niques render cancer operations much less morbid than in the past.

Increased understanding of cancer's presence, causes, morphology, pathogenesis and natural history is making this devastating disease seem more rational. It also raising hope that it can be entirely blocked, even before it occurs. As science labors to cure cancer, we should take heed from what we are learning about its causes. Important epidemiological factors have been identified that contribute to cancer. Foremost among these is smoking, which contributes to lung, esophageal, head and neck, bladder, pancreas, kidney, and other cancers. Other important social contributors to cancer risk include aging (prostate, gastrointestinal, other cancers), sun exposure (skin cancer, melanoma), sexual practices (cervi-cal, anal cancer), viral infections (cervical cancer, hepatocellular carci-noma, possibly Hodgkin's disease), diet, and obesity. It remains to be seen how readily we can alter societal habits to reduce cancer risk. One does not have to "cure" what never happens.

In the meantime prevention—through nutrition and other lifestyle choices—is our strongest combative measure. But it is up to each of us to shoulder that defense *personally*. One reason we do not hear more about prevention is money. No one makes money from prevention. Money also rears its ugly head in clinical trials. In the June 19, 2000 issue of *Business Week,* commentator Paul Raeburn reported that Dr. John Macdonald, director of St. Vincent's Comprehensive Cancer Center in New York City, took seven years just to enroll 600 patients for a study on stomach cancer. Stomach cancer affects about 22,000 Ameri-cans per year and is fatal about eighty-five percent of the time. So, if ever there was an experimental treatment one wanted to be enrolled in, this was it. And this prolonged time to complete a study that should be completed in a year is the rule not the exception. Why did it take Dr. Macdonald so long to register the needed patients?

Only two to three percent of all cancer patients participate in studies such as Dr. Macdonald's, and that is not because the current therapies are curing them. When surveyed most cancer patients did not even know that experimental treatments for their cancer were an option. Of those that did, many worried that insurance would not cover it, that the standard therapy was better, or that they would receive only a placebo. If the National Institutes of Health funded clinical trials of these treatments and educated patients about their benefits, instead of rat studies, more people would know about them and take advantage of them.

People must be educated to understand what their chance of survival is with whatever cancer they have. If they have a cancer with a ninety-five percent five-year survival or ninety percent ten-year survival, they may not wish to participate in an experimental study. But those whose doctors explain that their stomach cancer has less than a twenty percent five-year survival rate might be very pleased to hear that the NIH was sponsoring the study and that there would be no problem with insurance. These people would have a much better chance of survival and the entire field of oncology would accelerate if money went to projects like this instead of studying animal models.

Oncologists would also have to take the time to explain that cancer patients do not receive placebos. The control group receives the standard current therapy while the experimental group receives that plus the new treatment. But in order for a physician to participate in a clinical trial he must fill out much paperwork and confer many hours that he would otherwise not invest. So unless he is getting reimbursed for this, why should he do it? Change requires incentive . . . in the form of *money*. Only public pressure, roused by awareness over the failure of existing protocols, can force this revolution.

chapter 4

Development of Medications

For the great majority of disease entities, the animal models either do not exist or are really very poor. The chance is of overlooking useful drugs because they do not give a response to the animal models commonly used.

—Collin Dollery,
Hammersmith Hospital, London, in the book
Risk-Benefit Analysis in Drug Research

All the talk about genes notwithstanding, few people know what genes do, beyond establishing hair and eye color, height and so forth. Yet, a clearer acquaintance with genes is critical to understanding how drugs affect us. And it is central to how drugs will work in the future as researchers increasingly use genomic approaches to drug discovery and development.

Genes and Drugs

Genes direct the synthesis of proteins, which we mentioned in an earlier chapter. This is essential because our bodies are largely proteinaceous. According to instructions from genes, proteins occur in long chains of amino acids, called "sequences," that fold into distinct shapes.

Since there are thousands of different proteins, each functioning in a different capacity, one would imagine that many amino acids are involved. But no, it turns out that there are only around twenty amino acids. The number of each, their arrangement in the sequence, and the shape the chain folds into determine which protein is which.

Proteins are ubiquitous and essential to all plant and animal life. As hormones, they serve as the body's messengers. They act as gatekeepers in the form of receptors on the cell surface, delivering chemical messages to the cell's interior. These receptors also recognize harmful foreign matter and alert the immune system, activating yet another set of proteins, the antibodies. This protein function is important as pharmacologists are presently developing drugs to block receptor binding.

As it relates to pharmaceuticals, protein's most important aspect is as enzymes. Enzymes are our body's catalysts, regulating cell chemistry and controlling the metabolic byways through which more cells are manufactured. Pharmaceutical companies concentrate on a relatively small number of *drug-metabolizing enzymes*. These DMEs, as they are called, are responsible for metabolizing the majority of drug therapies in clinical use today.

One of the most major challenges in pharmaceutical development is figuring out how to deliver drugs through the cells' lipid (fatty) barrier to their correct destination inside the cell. The way chemicals signal the interior of the cell long remained enigmatic until recently when two scientists, Alfred Gilman and Martin Rodbell, determined the role of "G proteins" in cellular communications. Visualize a chemical arriving at the outer portion of the cell, the cell membrane, with a message to relay into the interior. These substances are either too large to gain entrance into the cell or are disallowed access for other reasons. Therefore, they only contact the cell on its surface. When they bind with the cell membrane, a body-manufactured substance called G protein, which sits in the cell wall, triggers a "second messenger," cyclic AMP (adenosine monophosphate). Cyclic AMP alerts the inside of the cell as to the message on the outside. Without G protein, the entire process does not work. For this exciting *in vitro* discovery, Gilman and Rodbell received the Nobel Prize, and the National Institutes of Health committed $25 million to unraveling more mysteries of cellular communication with Gilman at the helm.[1]

We now know that G protein is involved in over a third of all communications between cells and chemicals. It plays a role in glucose metabolism, stress response, muscle movement, fat metabolism, light sensing, odor sensing, and hormone response. If G proteins malfunction, disease ensues. Further, most medications probably act via G proteins.[2] Pharmaceutical companies want to study the drug-metabolizing enzymes, in addition to proteins that interact with G proteins since between the two, most diseases can be fought.

Each component of the cell has a job. Each performs these functions at the direction of proteins made both inside and outside the component. Sometimes a chemical in the cell cytoplasm activates the production of a protein. Further complicating our understanding of how medications might act on the cell is the fact that the components too are separated from the cytoplasm by a lipid barrier that does not normally allow proteins to pass.

The 1999 recipient of the Nobel Prize in Medicine or Physiology, Dr. Gunter Blobel, of Rockefeller University in New York, illuminated this problem when he characterized how the protein transport system works. Blobel discovered that proteins carry a variety of molecular signals that he called *zip codes*. These zip codes direct the proteins from the endoplasmic reticulum, where they are assembled in the cell, to the cytoplasm and into other cellular components or organelles, such as the nucleus. Blobel's was a classic *in vitro* discovery, as was Gilman and Rodbell's. This "signal hypothesis" is critical for several reasons. First, the transport system is universal, operating the same way in the cells of plants, yeast and animals. Second, mutations in one or more signals or mechanisms involved in the system cause hereditary disorders. Finally, the information enhances the efficiency of the use of cell culture systems for drug synthesis, which we will discuss in a moment. Dr. Blobel's work may make it possible ". . . to construct new drugs that are targeted to a particular organelle to correct a specific defect," according to a statement from the Nobel Assembly at Karolinska Institutet in Stockholm, Sweden. As gene technology evolves and the mapping of the human genome completed, the Assembly added, manipulation of the protein signals discovered by Dr. Blobel might even further enhance the use of cells as "protein factories."[3]

This is why genes are so important; they make sure organisms are equipped with the necessary proteins to communicate, and through communicating, maintain health. When these genes are mutated, and the mutations are expressed (meaning that the mutation is active rather than simply present), organisms either are prone to sickness or are sick in ways which medications may be able to alleviate.

In pharmaceutical development, scientists must design a drug to navigate a species that is essentially an incredibly complicated intake and output mechanism. Each continually filters, uses, and expels myriad substances—air, food, drugs, toxins, viruses, bacteria, and so forth. When any substance enters a body, a complicated series of actions determines its influence on that physiology. First it must be absorbed, either through respiration, the skin, or the gastrointestinal system. Once absorbed, the chemical courses its way into various tissues. Enzymes in the tissues then modify the chemical or allow the chemical to do what it is meant to do inside the cell. DMEs, directed by genes, determine what is metabolized and how, as just characterized. Ultimately, other enzymes connect the altered chemical to a large carrier molecule that escorts it from the body as sweat, urine or feces.

In the not-too-distant past, it was first the job of the medicine man and then that of the neighborhood pharmacist to procure therapeutic substances and deliver them to the ailing. As these people knew only that the substances worked, but not how, treatments were based on their authority rather than science. The era of modern-day drug discovery had to await advances in chemistry and physiology. Little by little science has taught us that although a certain remedy effectively alleviates disease, it is one particular component of the medication that is "active" or beneficial. Cranberry juice's effectiveness as a popular folk treatment for bacterial urinary infections is a good example. In the juice, scientists discovered chemicals called proanthocyanidins that keep the bacteria from sticking to the cells of the urinary tract.[4]

The next step, following the isolation of the chemicals, is their distillation into a more powerful medication. The first instance of this was in 1910 when Paul Ehrlich and Sahachiro Hata, learning that arsenic killed the syphilis-causing microorganism, synthesized arsphenamine, the first made-to-order drug to actually target and kill a disease-causing organism. Their compound initiated the pharmaceutical age. Over the last century, scientists continued to build their knowledge base about how chemicals, directed by enzymes, which are directed by genes, work. This knowledge informs pharmacologists' job today—designing substances that will interrupt the course of disease ever more precisely.

Where Drugs Come From

Animals need play no part in this design. The belief that drugs somehow spring from lab animals to our medicine chests is a complete myth. There are only four tried and true methods for finding fresh drugs:

- Discover new substances from nature as our ancestors did.
- Uncover a different curative value in an existing medication, as our more recent progenitors did and as often occurs in present day.
- Modify the chemical structure of a similar medication, either to improve on it or market yet another version.
- Design a new medication from scratch based on a desired action— this, the most thrilling and innovative alternative in modern pharmacology.

Cranberry juice reflects, of course, option number one. Nabilone is an example that falls somewhere between the first and second option. In the 1960s, after women reported that smoking marijuana reduced their morning sickness, pharmacologists developed nabilone from one of marijuana's chemical constituents.[5] It is now prescribed for acute morning sickness and for nausea ensuing from chemotherapy. But what happened when researchers introduced the animal element? Even though we already knew nabilone's positive effect in humans, researchers went ahead with animal experiments anyway. The results were neither good nor consistent. Nabilone produced toxic reaction in dogs but not rhesus monkeys or rats.[6] All the animal model did was waste money and time. Nabilone was already a safe, effective drug for humans. Another case in point is isoniazid, a valuable antituberculosis drug that is nontoxic to monkeys and humans, but that researchers needlessly proved kills dogs.[7] Of course, demonstrating these effects added nothing to the reservoir of information relevant to humans.

The food preservative nisin Z presages a new class of antibiotics. As an additive, nisin has been known for years to kill bacteria, but only with the advent of multidrug-resistant bacteria did researchers seek to find out why nisin worked. Drug discovery method number two. Via *in vitro* research, scientists found that nisin Z attacked bacteria at the same molecular location as does one of the most powerful antibiotics, vancomycin. The finding could lead to antibiotics more active than vancomycin by two orders of magnitude.[8]

Another example of method two is triclosan, an old antibacterial agent used in mouthwashes, antiacne agents, and deodorants. It is now under development as a treatment for malaria. The reason researchers believe that triclosan may work is that humans, animals, bacteria, plants—indeed all living things—have enzymes exclusive to them. That is why some parasiticides kill fleas without harming humans. Humans lack a protein that fleas have, and therefore the poison works only on the flea. The parasite that causes malaria has a unique enzyme, FabI, which helps synthesize fatty acids and build cell membranes. Triclosan is deleterious to this enzyme. This was an *in vitro* discovery. It has been tested on mice, which merely duplicated the *in vitro* findings. Not until human trials will researchers learn if humans can tolerate triclosan internally.[9]

In option number three, pharmacologists synthesize new chemicals based on existing medications, or on substances from nature that appear to have curative value. Afterward they hope that the new drug will work better and with fewer side effects than the original. There have been some successes and some failures. In the last thirty to forty years, biotechnology has emerged and flourished along lines described later in this chapter, eliminating a lot of the guesswork, thus allowing new drugs to issue from tried and true formulas.

The field of infectious disease offers interesting examples of a combination of all developmental approaches characterized above. Out of *in vitro* research emerged a new class of antibiotics known as oxazolidinones that are active against plant pathogens. *In vitro* testing of the chemicals against otherwise-resistant strains of bacteria that cause disease in humans found them effective.[10] Modification of the chemicals led to two antibiotics we have today, linezolid (Zyvox) and perezolid.[11] They are active against multidrug-resistant bacterial infections from *staphylococci, streptococci,* and *enterococci.*[12,13] All indices reflected the drugs' benefit to humans. Nevertheless, many animals were infected and given these new drugs, as required by law and by convention. This led to no new information. Patients who were dying from methicillin-resistant *Staphylococcus aureus* (MRSA) and other diseases could have received these chemicals instead. They might have lived, and the data garnered would have had human value. (Interestingly, with linezolid the animal tests did not reveal a myelosuppression side effect. Had more extensive clinical trials or more extensive *in vitro* testing been performed, this might have been noticed prior to the drug being released to the general public. In 2001, the manufacturer Pharmacia relabeled the drug accordingly.)

Since the 1960s, the pharmaceutical industry has been using computers and models to analyze the relationship between given chemicals' structure and activity. This analysis supplies an accurate picture of life processes and the forces that act on them at the molecular level. The methodology, known as *quantitative structure-activity relationships* (QSARs), defines the relationship of a chemical's biological activity to its atomic arrangement, electrical charge, and other properties. In other words, it shows the chemical's molecules in action. Obviously, these models are a most useful venue for establishing where and how drugs must operate. Structure-function analysis or structure-activity analysis— the fourth option of those enumerated—is how most drugs have been designed for the last half century.

Combinatorial chemistry was born in the 1980s as a result of *in vitro* research on solid-phase synthesis of oligonucleotides, (polymers made up of between two and twenty nucleotides) and peptides (the constituent parts of proteins). Enhanced by the marriage of *in vitro* research with computer technology, combinatorial chemistry facilitated the synthesis of thousands of chemical variants of a known therapeutic drug.

Computer-aided drug design (CADD) programs cull and integrate results from many *in vitro* research efforts. A four-step process, it is the perfect tool for creating chemicals and analyzing their effect on human cells. The four steps are:

- By watching how disease enters the cell *in vitro*, scientists target the cell receptor they wish to activate or deactivate, then they program in the receptors' characteristics on the computer.
- Scientists identify candidate medications whose structure suggests that they might activate or deactivate given cell-receptors. (This process involves a process called *high-throughput screening*.)
- Scientists submit the chemicals to the cells to see if they work. (This process is called *high-content screening*.) Using a computer, scientists model the receptor they wish the chemical to bind with (for example, a receptor on the surface of CD4 cells that would block the entrance of HIV into the cell) and the

drug that they believe will accomplish this. High-content screening evaluates their physiological properties by combining fluorescence-based reagents with hundreds of cells located in small wells on a plate. Each well yields a positive or negative reaction, thus informing the researcher of what the potential medication really does in the cell.

- Based on this data, scientists pick the best chemical for the job.

Structure-activity analysis has been used for decades but advances in technology and chemistry have allowed it to really take off in the last decade. Advances in technology allow the pharmaceutical industry to test greater than 160,000 chemicals over a few months. Computers evaluate chemical properties and interactions via robotics using miniature, mechanized methods. [14]

Computers contribute to drug design in two basic ways. When the shape of the receptor the drug must bind to is unknown, they allowing scientists to conceptualize the receptor's conceivable shape. If there are five possibilities and a certain drug works with four, they may opt for that drug. If the shape is known, computers can design a drug around the shape. Protease inhibitors, which curtail one of the enzymes that HIV requires to reproduce, were some of the first drugs designed by computer.[15] You can see how efficiently computers and *in vitro* analysis let scientists discern a drug's efficacy before it ever is given to humans. Researchers can work it both ways, either comparing computer-synthesized results with *in vitro* activity or vice versa. Both modalities indicate if the drug displaces the molecules that bind to a known active receptor.[16] These methods are certainly superior to the highly unreliable use of animal models.

Those with a vested interest in animal models cannot let superior technology disturb their myth. They want to keep the public thinking that rodents, dogs and primates are still vital to drug development and safety. But despite their efforts, advances in technology have made animal models more obsolete than ever. The richness of our intelligence about catalytic processes on a molecular level, and about the molecular causes of disease continues to supersede the animal model. Through technology combined with *in vitro* research, scientists can isolate the molecules and purify them to determine their three-dimensional structure at atomic resolution. Then, they can monitor to prove or disprove that the substances act as anticipated at a cellular level. Instead of testing on animals, whose genes, metabolism, and habits are much different from humans, scientists can impregnate a silicon chip of two to three centimeters in width with strands of human DNA, then observe directly if a new potential drug triggers a gene to act, or stops it from acting.[17]

Few drugs turn out to be one hundred percent effective for all humans, for good reason. Response to drugs differs according to environmental and genetic factors. Environmental factors include smoking, alcohol and drug use, diet, occupational exposure to chemicals, and disease. Each of these can impact the way certain drugs operate within the body. In addition, minor variations in genes often sway responses to the same drug. These "genetic polymorphisms," as they are known, may reduce a drug's efficacy and/or result in toxicity. When inheritance varies response to drugs it is called *pharmacogenetics.*

Even though all human genomes are greater than 99.9 percent identical, the small 0.1 percent difference predicts as many as three million polymorphisms. Many polymorphisms in the 40,000 or so genes in the human genome have no effect. Some, however, impact protein expression and function, putting response to specific medicines off-kilter. When a drug is designed to affect a certain protein and the protein in question is polymorphic or different, the drug does not work as designed in certain individuals. For instance, polymorphism in the cholesterol ester-transfer protein (CETP) impedes efficacy of the drug pravastatin in patients diagnosed with coronary atherosclerosis. Polymorphisms in beta-adrenergic receptors affect sensitivity to beta-agonists such as albuterol taken by asthma patients. And polymorphisms in the serotonin neurotransmitter receptor (5HT2A) diminish the effectiveness of the antipsychotic drug clozapine.[18]

Sometimes, genetic polymorphism in the DMEs themselves contorts the effectiveness of the drugs. There exist three distinct subgroups of people who have measurable pharmacogenetic differences in their ability to metabolize drugs to either inactive or active metabolites. Individuals capable of efficient drug metabolism are called *extensive metabolizers* (EMs); most people fall into this category. Individuals with deficiencies in metabolism, due to gene deletion, are termed *poor metabolizers* (PMs). Conversely, over-expression due to gene amplification results in *ultra-rapid metabolizers* (UMs). Standard doses of drugs with a steep dose-response curve or a narrow therapeutic range may produce adverse drug reactions, toxicity, or decreased efficacy in PMs. When taken by UMs, these standard doses may be inadequate to produce the desired effect.

Polymorphisms in DMEs that bear considerably on clinical medicine include those affecting the cytochrome P-450 liver-enzyme family and the enzyme thiopurine methyltransferase (TPMT). These genetic variations, apparent in a significant percentage of the population, diminish the therapeutic efficiency of drugs commonly used to treat cardiovascular disease, cancer, central nervous system disorders, and pain. Approximately one out of ten people has genetic polymorphism in the gene that expresses the P-450 enzyme, and the percentage of people with genetic polymorphism in TPMT is the same. The P-450 bears on approximately twenty-five percent of all drugs' metabolism. That means that for one-quarter of all drugs, ten percent of the people who take them will not benefit as intended. That is a significant degree of failure. Within the last ten years, it was estimated that more than 2,200,000 hospitalized patients had serious adverse drug reactions and 106,000 had fatal adverse drug reactions.[19] A considerable portion of these reactions is due to genetic polymorphism. Genetic tests now exist to predict polymorphism. The degree of variability these tests determine even between individual humans, that is members of the same species, is sobering evidence that animal experiments are more clumsy and inaccurate than ever.

As government-funded and private efforts expand the knowledge of genes and the proteins they govern, the goal of achieving one-hundred percent effectiveness with drugs seems within reach. This is an anticipated outcome of the mapping of the human genome and the subsequent understanding of the proteins for which the genes code. The journal *Science* stated:

> The Human Genome Project, coupled with functional genomics and high-throughput screening methods, is providing powerful new tools for elucidating polygenic components of human health and disease.[20]

The objective is to someday modify or customize drugs for each patient's chemistry, producing therapies that are tailor-made adaptations to environmental and genetic variations. Gene-tailored drug discovery, still in incipient phases, is called *pharmacogenomics*.[21] This intelligence of the poly-genes—multiple genetic influences on disease—is beginning to support the manufacture of medications specifically designed for individual genomes, to treat cardiovascular diseases, cancer, bone disease and others as they uniquely manifest.[22]

Once a gene or genes are identified as involved in a particular disease, scientists can test how a potential drug may affect them by using DNA microarray technology. The drug may turn the gene off or on and this will show up as a pattern of fluorescent spots on the gene chip, as explained earlier.

Herceptin, one of the first approved drugs to issue from pharmacogenomics, inhibits the *HER2* oncogene (a cancer-causing gene). Approximately twenty-five to thirty percent of breast cancer patients over-express the *HER2;* the over-expression leads to an increase in the replication of the cancer cells. Doctors can identify which patients over-express *HER2,* and then administer the drug to them. Instead of giving large doses of chemotherapeutics and killing all cells, the drug targets only the cells that need to be killed. Those who will not benefit from the drug need not take it.

Herceptin is a monoclonal antibody designed to act on a specific destination. One of several of a new class of drugs called epidermal growth factor (EGF) receptor blockers, Herceptin binds to receptors on the cell surface and blocking them from receiving signals that direct the cell to divide and become cancerous. These stabilize rather than shrink tumors. Researchers anticipate that these new treatments will someday transform cancer into a chronic condition that can be kept in check.[23]

Herceptin is also interesting because it improves on initial monoclonal antibody disasters. In the 1980s monoclonal antibodies were considered a promising area of exploration for cancer treatment. Scientists fabricated them by injecting mice with chemicals that would then lead the mice to produce the antibody. However, when researchers injected antibodies that came from mice into human subjects, small pieces of protein molecules that came with the antibodies from the mice produced an immune reaction. This rendered the treatment problematic. Research lagged after that—another delay due to animal use. Herceptin was made by substituting human proteins for the mouse proteins. Monoclonal antibodies are now manufactured without animals (as the very first ones were), thus obviating the problem altogether.[24]

Authorities anticipate that by 2005 genotype testing will be a routine procedure before prescribing many drugs. As one scientist summed up at the Millennial World Congress of Pharmaceutical Sciences, "Such an approach promises to identify novel individual predictors of drug response and should greatly aid in obtaining the goal of prescribing prescriptions with a personal touch."[25] In the future, it will be considered unethical to expose patients to the risks of adverse events without first performing these fast, simple DNA tests. Improving patient outcomes and avoiding adverse reactions will reduce the costs of hospitalization, the number of office visits, and the waste incurred by ineffective therapy.

Drug Testing with the Animal Model

The intake-output transport function described a few pages ago is a very complex business, and is one reason why drug development takes time. To be successful, any drug must be at once effective and not deleterious to patients' health. When chemicals are introduced into the body, it is possible that they will do more than interrupt the disease-causing mechanism; there is a chance they will also damage parts of the body.

Therefore, drugs undergo three basic testing phases before being released to the general population. First is the *in vitro* research and testing sometimes accompanied by computer modeling and mathematical modeling that we have already outlined. The second is the animal testing phase and third the clinical trials phase in which the drug is tested *in vivo* on humans. But before a new medication even reaches clinical trials, drug companies must create what they call a *pharmacokinetic profile*. This profile essentially attempts to anticipate the drug's half-life, its distribution, its interaction with other drugs, whether or not it damages cells (cytotoxicity), and whether or not it damages the DNA (genotoxicity). If the pharmacokinetic profile shows persuasive evidence that the drug is effective and safe, the substance proceeds to trials on humans.

The traditional approach to profiling, in place since the 1960s, includes animal models. As explained earlier in *Sacred Cows and Golden Geese,* animal testing became mandatory during the 1960s, and today, though it is clearly not a reliable predictor of human response, the convention is still used as a fail-safe against litigation. Drug companies conduct a lot of animal tests so that, in case the drug makes people sick or kills them, they can later point to the rigorous animal tests that they performed and claim that they did their best to ensure against such tragedy, thus minimizing the monetary judgment against them.

The courts accept this indemnification despite countless examples illustrating the ineptitude of animal-modeled methods. One has only to look at the many meritorious drugs that went directly from discovery to humans without animal testing before it became requisite—ipecac, aspirin, cinchona bark, digitalis, and the early inhalation anesthetics to name but a few—to recognize the dissimilitude between animals and humans.[26] Had the requirement for animal testing then existed, these valuable drugs would never have made it to the market. We know this because researchers eventually tried them on animals, despite already knowing that they would work on humans, and found hazardous side effects.

The development of antibiotics is a watershed example. In 1877, Pasteur and Joubert observed that a microorganism's contaminants inhibited the growth of *Bacillus anthracis.* This led others to experiment with such interactions between microbial species. More than fifty years later (years heavily influenced by animal experimentation), Sir Alexander Fleming serendipitously witnessed the inhibition of colonies of *Staphylococcus aureus* by a contaminant from the mold *Penicillium* in 1928.[27] Fleming tried his "penicillin" on a rabbit but abandoned it because it was ineffective. A decade later, with a quickly deteriorating human patient, he tried it again out of desperation. And antibiotics were born.

Fleming might have thrown his penicillin away entirely if he had tried it first on guinea pigs or hamsters. Penicillin makes these species bleed from the intestines (due to penicillin's effect on the gut bacteria on which species with a fore-gut rely to break down cellulose) and kills them.[28] Penicillin is teratogenic (causes birth defects) in rats, causing limb malformations in offspring.[29,30,31] Fleming himself told one of his students, "How fortunate we didn't have these animal tests in the 1940s, for penicillin would probably never been granted a license, and possibly the whole field of antibiotics might never have been realized."[32] One really cannot disparage Fleming for his reliance on animal testing since in his period the perceived similarities between species outweighed the apparent differences by many times. But today we have better methods. Researchers study disease and treatment at the very level that differentiates species—the molecular level—as described in the previous chapters.

The tragedy of thalidomide, an incident that coincided with the legislation requiring animal testing, is another example. As covered in detail in *Sacred Cows and Golden Geese*, rats did not predict that thalidomide would cause phocomelia (deformation of extremities) in offspring.[33,34] Thalidomide (N-phthalidomido-glutarimide) is similar in structure to two sedatives introduced in Germany in the early 1950s—diazepam, better known as Valium, and barbital. Because of the similarity, a German company called Chemie Grünenthal first manufactured and sold thalidomide as a sedative in 1957. Chemie Grünenthal tested thalidomide on pregnant humans in countries other than Germany and saw no problems—none in the first trimester of their pregnancy that is. Testing thalidomide on humans in other countries was unethical but not surprising since the medical director at Grünenthal, Heinrich Mückter, had been involved in experiments on civilians at Krakow, for the Nazis during World War II. Was thalidomide tested on pregnant animals in Germany? Yes.[35] Specific teratogenicity testing may or may not have been done, but general animal tests certainly were. Roald Hoffmann writes in *The Same and Not the Same*:

> Indeed animal testing for teratogenicity of new drugs was routine in the major pharmaceutical companies. Hoffmann-LaRoche's Roche Laboratories published a major reproductive-system study of its Librium in 1959. Wallace Laboratories did so for Miltown in 1954. Both incidents antedate the thalidomide story.[36]

When reports of birth defects from thalidomide began appearing, Grünenthal stepped up advertising campaigns and threatened physicians who reported that thalidomide was dangerous. After thousands of malformed human babies were born, all of mothers who had taken thalidomide, and after researchers had failed to produce similar malformations in numerable other species, they finally found that an obscure breed, the White New Zealand rabbit, also gave birth to malformed offspring, but only at a dose between 25 to 300 times that given to humans. Eventually some monkeys gave birth to deformed offspring too, but it took ten times the normal dose to make this happen.[37,38,39] The bottom line is this: more animal testing would not have prevented the release of thalidomide, because scientists would not have found the side effects. Even if they had tested the White New Zealand rabbit, thalidomide would have still come to market since the vast majority of species showed no ill effect from the drug. This following paragraph from a government-issued statement is not the only publication to have stated as much.

> There is at present no hard evidence to show the value of more extensive and more prolonged laboratory testing as a method of reducing eventual risk in human patients. In other words the predictive value of studies carried out in animals is uncertain. The statutory bodies such as the Committee on Safety of Medicines [Britain's counterpart to the FDA] that require these tests do so largely as an act of faith rather than on hard scientific grounds. With thalidomide, for example, it is only possible to produce specific deformities in a very small number of species of animal. In this particular case, therefore, it is unlikely that specific tests in pregnant animals would have given the necessary warning: the right species would probably never have been used. [40]

Vested-interest groups such as the Foundation for Biomedical Research and Americans for Medical Progress perpetuate the myth that the United States government did not approve thalidomide because animal tests had raised suspicions about the drug. The facts are different. Frances Kelsey, a medical officer at the FDA stated the decision not to allow thalidomide was based on the fact that it led to peripheral neuritis, numb and tingling fingers in adult humans.[41] Animal tests had nothing to do with the decision. (The use of thalidomide is making a comeback, again based on human-based research. In 1964 an Israeli physician prescribed thalidomide as a sedative for a patient suffering from leprosy. He noticed it actually alleviated the leprosy. Some cancer patients also benefit from thalidomide.) Be all this as it may be, the thalidomide epoch initiated the requirement for animal testing of prospective drugs.

Despite the animal model's acknowledged inefficacy, pharmaceutical companies, vigilant in their efforts to avoid lawsuits, still rigorously conduct animal testing to produce the pharmacokinetic profile. To calculate human values, scientists scale the dosage to animal models by weight and make what other adjustments they can in order to stretch what is happening in the surrogate into human terms. Professor Andre McLean, Department of Clinical Pharmacology, University College, London, speaking at a conference reported in *Animals and Alternatives in Toxicology,* stated:

> Yes, I think it is very clear to all of us who are engaged in the business of assessing toxicity data that, when volumes of data are proudly presented to us after a carcinogenicity study [on animals] showing that there was a tumour in this organ or that, we look at it and we scratch our heads, and we wonder what on earth we can make of it. This is especially true when huge doses are given, with nothing to suggest what would be expected at low doses. I think

very often the carcinogenicity studies are a waste of everybody's time and a fearful waste of animals. They are conducted partly because we are not sure what to do instead, and partly because they are a political gesture and a very miserable one at that.[42]

To give you an idea of how challenging the extrapolations are, consider the subject of *protein binding*. When a drug is administered, a portion of it attaches to proteins in the blood. That portion is therefore not available to get into tissues and act. This protein binding, as it is called, in part determines the quantity of drug one has to take in order to see an effect. In a comparison of the protein binding of twenty-two drugs in rats and humans, one study found thirteen drugs bound less in the rat while seven bound more, and two were equally bound in both species. In comparing protein binding of fourteen drugs in humans and monkeys, seven bound more and seven less in the human than the monkey. Weighting the dosages will not adjust for these unpredictable incongruities. Such is the case of animal models. [43] For this reason, the words of Dr. John Griffin, Director of the Association of British Pharmaceutical Industry, "[Animal carcinogenicity tests on new drugs are] inaccurate, often insensitive and generally misleading" are often repeated.[44]

The only way to embrace research with this many nonquantifiable aspects is simply to discount the incongruities, obviously not a very scientific approach. This renders the science of toxicology "terribly imprecise in relation to drug development," as wrote one toxicologist. "Drug testing toxicology seems to have developed haphazardly and many of the guidelines that both the CS [Committee on Safety of Medicines] and the industry keep do not appear to have a very firm base."[45]

"As a very approximate estimate, for any individual drug, up to twenty-five percent of the toxic effects observed in animal studies might be expected to occur as adverse reactions in man," wrote another esteemed scientist.[46] But which effects constitute the twenty-five percent? Random chance can sometimes do better than a twenty-five percent correct prediction rate. As more than one authority has pointed out:

> Unfortunately, this use of animal models for predicting risks for man is beset with difficulties, it is rarely possible to be sure that the animal model properly represents the relationship in man. . . . Even if it were possible to improve the accuracy of the present-day test procedures, so that the risk to the test animals were known with greater precision, this would not necessarily bring about a corresponding improvement in the assessment of potential risk to man, because of the uncertainty regarding the relevance of the animal data for man.[47]

The animal-model method just shrugs aside fundamental differences, and because it does so, the results are unreliable. Yes, animals are "intact systems" that have features in common with man. However, as the book *Safety Testing of New Drugs* put it:

> Differences in tolerance between species, and between strains within species, differences in metabolism between species and between strains, differences in lifespan, and in duration of dosage can all be argued to make a given result crucial, or negligible, according with the preferred point of view. In such circumstances the comfort given by decisions based on an extended range of [animal] tests is largely an illusion.[48]

As Dr. Bjorn Ekwall, chairman of the Cytotoxicology Laboratory (CTLU), based in Sweden said:

> Prediction of human lethal and toxic doses is poor due to species differences between animals and humans, and the toxic mechanisms of the chemicals cannot be directly predicted using current animal tests.[49]

By design, animal testing is supposed to *predict* efficacy and safety. Yet, comparisons of side effects are so *inconsistent* that the convention can only be deemed entirely nonpredictive. As an example, researchers took six drugs with known side effects in humans. Animal testing correctly predicted twenty-two side effects, but incorrectly presented forty-eight side effects that did not in fact occur in humans. Further, the animals did not predict twenty side effects that did occur in humans. Thus the animal models were in error sixty-eight out of ninety times. In these incidences, seventy-six percent of the time results from animal experimentation were wrong.[50]

To cite a specific drug, take the case of fenclozic acid, which was a potential new anti-inflammatory. No adverse side effects were observed to occur in the mouse, rat, dog, rhesus monkey, patas monkey, rabbit, guinea pig, ferret, cat, pig, cow, or horse, yet the drug caused acute cholestatic jaundice in humans.[51]

Pharmaceutical companies perform comprehensive animal tests, putatively to protect humans during the clinical testing phases as well as end consumers, but their actual objective is just to get the substance onto pharmacy shelves and to protect themselves legally if a problem arises. That humans are not at all protected by these animal tests is evident in all the illness and mortality that occur during the clinical trial phase. For instance, 122 deaths were associated with Lotrafiban, a drug designed to fight strokes and heart attacks before the trials were halted.[52,53] So much for animal testing to protect those undergoing clinical trials.

Regarding toxicity trials for new drugs in humans and the disasters animal testing has precipitated, an unnamed clinician quoted in *Science* stated, "If you were to look in [a big company's] files for testing small-molecule drugs you'd find hundreds of deaths." The squandered time and resources that go into this sort of medication development would matter less if pharmaceutical companies would subject the substances to a subsequent more careful, more stringent level of human-based testing. But they do not. They want that drug on the market, earning money. As a consequence, animal-modeled toxicity testing has led to many, many drugs that, once tried on humans, are either recalled or relabeled. (Drugs are relabeled when a severe side effect occurs and the FDA or manufacturer believes that the public should be especially aware of this risk before taking the drug. For instance, if a medication designed to be taken for mental depression causes bone marrow depression. The patient, along with his physician, may decide to take a different drug or may decide the drug is still worth taking. But, in either case, they would be making an informed decision.) Yes, drugs that are developed around animal models—animals that benefit from the drug and show no adverse reactions in toxicity testing—frequently prove deleterious to humans. Moreover, it is usual that they are not recalled until they have already garnered substantial profit. As R. W. Smithells wrote in 1980 in *Monitoring for Drug Safety*:

> The extensive animal reproductive studies to which all new drugs are now subjected are more in the nature of a public relations exercise than a serious contribution to drug safety.[54]

The examples of these mistakes are almost innumerable. The list below, from the United Kingdom, further illustrates the point. Between 1980 and 1986, in the UK, the following drugs were withdrawn for the following reasons:

- Phenacetin, an analgesic—carcinogenicity and kidney toxicity. Withdrawn 1980.
- Clioquinol, an antidiarrheal—neurotoxicity called SMON syndrome. Withdrawn 1981.
- Benoxaprofen (Opren), an anti-inflammatory—a variety of adverse effects, including deaths. Withdrawn 1982.
- Phenphormin, an antidiabetic—metabolic toxicity. Withdrawn 1983.
- Zomepirac (Zomax)—allergic reactions. Withdrawn 1983.
- Zimeldine (Zelmid), an antidepressant—neurotoxicity. Withdrawn 1982.
- Indomethacin-R (Osmosin), an analgesic—gastrointestinal toxicity. Withdrawn 1983.
- Indoprofen (Flosint), an analgesic—gastrointestinal toxicity. Withdrawn 1983.

- Propanidid—allergies. Withdrawn 1983.
- Feprazone (Methazone), treatment for rheumatic and rheumatoid arthritis as well as osteoarthritis—life-threatening blood abnormalities. Withdrawn 1984.
- Fenclofenac (Flenac)—gastrointestinal toxicity and carcinogenicity. Withdrawn 1984.
- Alphaxalone, an anesthetic—anaphylactic shock. Withdrawn 1984.
- Nomifensine (Merital)—immune and blood toxicity. Withdrawn 1986.
- Domperidone, an antiemetic—cardiotoxicity. Withdrawn 1986.
- Guanethidine, antihypertensive eyedrops—ophthalmologic toxicity. Withdrawn 1986.
- Sulfamethoxypyridazine, an antibiotic—blood toxicity. Withdrawn 1986.
- Oxypenbutazone, analgesic, anti-inflammatory and anti-arthritic drug—blood abnormalities. Withdrawn 1986.
- Suprofen (Suprol) an analgesic—kidney toxicity. Withdrawn 1986. [55]

In the United States, the FDA cited the following pharmaceuticals:

- Raplon (rapacuronium bromide), a muscle relaxant for surgical procedure—five deaths from bronchospasm. Withdrawn 2001.
- RotaShield, rotavirus vaccine—intussusception and death. Withdrawn 2000.
- Phenylpropanolamine, an ephedra alkaloid, a cold and decongestant medication, appetite suppressant—cerebrovascular events such as hemorrhagic stroke.[56] Withdrawn in 2000.
- Basiliximab (Simulect) prophylaxis of acute organ rejection in patients receiving renal transplantation—acute severe hypersensitivity reactions. Relabeled in 2000.[57]
- Raxar (grepafloxacin), a fluoroquinolone antibiotic. Withdrawn in the 1990s it caused heartbeat abnormalities, known as QT interval prolongation on the electrocardiogram. These can lead to a life-threatening ventricular arrhythmia known as torsade de pointes.[58]
- Phenacetin, analgesic—carcinogenicity and kidney toxicity. Withdrawn in 1980.
- Zomepirac (Zomax)—allergic reactions. Withdrawn in 1983.
- Sumatriptan, a headache medication—caused heart attacks and deaths by constricting the coronary arteries of healthy people and people with heart disease.[59] Relabeled.
- Serzone, an antidepressant—caused 18 cases of liver failure, leading in some cases to transplantation and death. Relabeled.[60]

- Zonegran (zonisamide), a new anti-seizure medication—toxic epidermal necrolysis (people sloughed their skin), psychosis and other side effects not predicted by animals.[61] Relabeled.
- Moxonidine (Moxcon), an antihypertensive medication used in end-stage congestive heart failure—deaths. Halted clinical trials.
- Alphaxalone, an anesthetic—anaphylactic shock. Withdrawn in 1984.
- Propanidid, an anesthetic—allergies and severe hemodynamic side effects. Withdrawn in 1983.
- Clioquinol, an antidiarreahal—neurotoxicity called SMON syndrome. Withdrawn in 1981.
- Phenphormin, an antidiabetic—metabolic toxicity. Withdrawn in 1983.
- Domperidone, an antiemetic—cardiotoxicity. Withdrawn in 1986.
- Guanethidine antihypertensive eyedrops—ophthalmologic toxicity. Withdrawn 1986.
- Sulfamethoxypyridazine, an antibiotic—skin and blood toxicity. Withdrawn in 1986.[62]
- Idoxuridine, an antiviral—ineffective and also fatal in some humans.[63,64] Withdrawn 1986.
- Lotronex, a drug for treating irritable bowel syndrome—linked to five deaths, the removal of a patient's colon and other bowel surgeries. Withdrawn in 2000.[65]
- Redux (grepafloxacin), a diet pill—caused heart-valve damage and respiratory failure, suspected in 123 deaths. Labeling made no mention of dosage data or deaths. Inexplicably, FDA approved despite unwillingness of examiners to sign off on it. Withdrawn in 1997.[66]
- Posicor, blood pressure medication—disrupts heart rhythm and interacts negatively with certain other drugs, suspected cause of 143 deaths. Scores of other drugs for treating high blood pressure were already on the market, and though Posicor did not offer other lifesaving benefit, the FDA approved it. Withdrawn in 1998.[67]
- Duract, a painkiller—suspect in 68 deaths, including 17 involving liver failure. Withdrawn in 1998 after generating almost 90 million in sales.[68]
- Propulsid, heartburn drug—caused heart-rhythm disorders in children as well as adults, cited as a suspect in 302 deaths. The problem had been identified as early as 1995, by the manufacturer and FDA kept it on the market with re-labeling until 2000. More children died.[69]

- Rezulin, a diabetes treatment—caused liver failure and death. Withdrawn 1999.
- Sporanox (itraconazole), an anti-fungal—caused congestive heart failure and death. Janssen Pharmaceutical Products, a unit of Johnson & Johnson, warned doctors not to prescribe to patients at risk of suffering heart failure in 2001.

The drugs enumerated here are just a smattering of the total calamities. Total reports of adverse drug reactions filed to the FDA by health professionals, consumers and drug manufacturers increased eighty-nine percent from 1993 through 1999. More than 250,000 side effects linked to prescription drugs, including injuries and deaths, are reported each year.[70] The reports are only those filed voluntarily and do not include adverse events that go unreported. Neither laws nor financial incentives encourage physicians to report.

Reliance on animal experimentation helps create these tragedies, and our government's new disposition toward the pharmaceutical industry increases the jeopardy to consumers. In the nineties, pharmaceutical companies put pressure on the government to bring all drugs to market more quickly, pressure in the form of lobbying and campaign contributions. And the government obliged. The FDA's approval rate went from sixty percent at the beginning of the decade to eighty percent presently. In 1988, the FDA approved only four percent of new drugs introduced into the world market. In 1998, the FDA's first-in-the-world approvals spiked to sixty-six percent.

What this means is mammoth domestic sales, which would be great if the drugs were needed to save lives. Seven drugs described in a *Los Angeles Times* article on the subject were withdrawn, cited as suspects in over a thousand deaths. Of these, six were never proved to offer lifesaving benefits and the seventh, an antibiotic, was unnecessary because other, safer antibiotics were available.[71] The article quoted a University of Texas School of Public Health physician who served from 1995 to 1999 on an FDA advisory committee, Dr. Lemuel A. Moye, who described FDA officials:

> They've lost their compass and they forget who it is that they are ultimately serving. Unfortunately, the public pays for this, because the public believes that the FDA is watching the door, that they are the sentry.[72]

No point in doubting the clout of the $100-billion pharmaceutical industry. In the last ten years, drug companies have steered $44 million in contributions to the major political parties and candidates. What this means is that the FDA is effectively financed by the pharmaceutical industry. The agency "works for" the industry, not for consumers. Why? Because consumers are not making the big campaign contributions.

The approval process for a putative flu medication, Relenza, is not atypical. FDA reviewers who examined Glaxo's drug found that Relenza was no more effective than a placebo in treating common flu symptoms and was potentially unsafe for flu patients with asthma or other respiratory disease. The agency, under pressure from the pharmaceutical company, took the chief reviewer off the case and approved it. Almost immediately, some patients taking the drug showed adverse effects to the extent that they were hospitalized and among them there were fatalities. Glaxo says it is "difficult to determine" whether Relenza caused the deaths. It is still on the market.

Animal testing, in a sense, cloaks all prospective substances; it gives them a double identity under which they can operate until their function or dysfunction in human application reveals itself. Positive animal tests also keep faulty drugs on the market longer than they should be. As Gina Kolata reported in the *New York Times* September 16, 1997, on the diet drug fen-phen:

> Why weren't these problems noticed before? Dieters in Europe had used Dexfenfluramine for decades. Dr. Friedman [an FDA official] said he could only speculate. No one had initially thought to examine patients' hearts, he said, because animal studies had never revealed heart abnormalities and heart valve defects are not normally associated with drug use.

Practolol (Eraldin), a heart treatment, was another potential medication that greatly surprised its developers. It caused blindness and twenty-three deaths despite the fact that no untoward effects could be shown in animals. In fact when it was introduced it was, ". . . particularly notable for the thoroughness with which its toxicity was studied in animals, to the satisfaction of the regulatory authorities." Even laboring long and hard after the drug was withdrawn, scientists failed to reproduce these results in animals. Echoes of thalidomide. As a government publication summed up, the practolol adverse reactions have not been reproducible in any species of animal except man.[73]

According to the *Los Angeles Times*,[74] Rezulin, the antidiabetes drug, is a glaring example of the FDA's "fast-track" approval process in action. In development, Rezulin lowered blood sugar in rats without hurting them. Warner-Lambert ran with that. But according to articles in the *Los Angeles Times,* Warner-Lambert also knew, based on human-derived data, that Rezulin could compromise the human liver well before the approval process began. Prior to submitting Rezulin for FDA review, Warner-Lambert removed a recommendation for monitoring the liver from the labeling. Had the recommendation remained, it would have slowed the drug-approval process. Company press and officials also claimed that the drug was the first anti-diabetic drug to target insulin resistance, and that it was virtually free of side effects.

Soon after Warner-Lambert submitted for FDA review in the summer of 1996, Dr. John L. Gueriguian, the medical officer assigned to examine it, cited Rezulin's potential to harm the liver and the heart. Gueriguian also questioned its viability in lowering blood sugar for patients with adult-onset diabetes. "You can't shine shit with words," Gueriguian said he told a Warner-Lambert executive. Since Warner-Lambert's mission statement is "Every day a new product fails to reach a market means missed opportunity," executives there were seemingly unconcerned about impediments such as harm and ineffectiveness. Dr. Murray M. "Mac" Lumpkin (now being considered for appointment as FDA commissioner) and Dr. Henry G. Bone III, chairman of the FDA advisory committee quietly collaborated with Warner-Lambert, according to the *Los Angeles Times*. Under pressure from Warner-Lambert, Gueriguian was stripped of the assignment in November 1996. The FDA staff withheld the existence of Gueriguian's review from the advisory committee that would decide on the drug, Warner-Lambert officials did not reveal that liver injuries in patients taking Rezulin were nearly four times as likely as for those given placebos, and the FDA approved it in January 1997.

Liver failures occurred immediately. Warner-Lambert launched a nationwide marketing campaign that included sales training seminars with hired actors playing physicians. The company became involved in the National Institutes of Health's nationwide diabetes study. Leading the $150 million study was Dr. Richard C. Eastman, who also served as a paid consultant to pharmaceutical companies. Eastman received $78,455 in 1997 from Warner-Lambert alone. The manufacturer also generated funding or compensation for at least twelve of the twenty-two scientists selected for the NIH study. Seven obtained up to $300,000 in grants, speakers' fees of $1,000 per address, and other stipends. So researchers, who were perforce biased, enjoined physicians nationwide to prescribe this dangerous and not necessarily effective drug as part of the study. Coincidentally, Warner-Lambert awarded funding that could total fifty million dollars to Dr. Jerrold M. Olefsky's team at the University of California, San Diego. Olefsky is an inventor of Rezulin. He is also a founding co-chairman of the National Diabetes Education Initiative, a group established and financed by Warner-Lambert that courted physicians to prescribe Rezulin. As a commodity, Rezulin was performing as the Warner-Lambert mission demanded—"not missing an opportunity to reach the market."

Meanwhile, back at the FDA, reexamination of the drug was sluggish. On the eve of a hearing to reassess the diabetes pill, the FDA appointed two new members to the advisory panel. Both had received income during the last two years as leaders of a diabetes-education group funded exclusively by Warner-Lambert and its Japanese partner. Though Rezulin killed hundreds of people and obliged hundreds more to undergo liver transplants, Warner-Lambert made over two billion dollars before it was withdrawn in 2000. [75,76]

But Rezulin did work beautifully in rats.

Summarizing the challenges of the animal-model method as it now stands, Dr. Anthony Dayan of the Wellcome Research Laboratories said:

> The weakness and intellectual poverty of a naive trust in animal tests may be shown in several ways, e.g., the humiliatingly large number of medicines discovered only by serendipitous observation in man (ranging from diuretics to antidepressants), or by astute analysis of deliberate or accidental poisoning, the notorious examples of valuable medicines which have seemingly 'unacceptable' toxicity in animals, e.g., griseofulvin producing tumours and furosemide causing hepatic necrosis in mice, the stimulant action of morphine in cats, and such instances of unpredicted toxicity in man such as the production of pulmonary hypertension by Aminorex and SMON. The rapidly increasing interest in clinical pharmacology, and the drive to better means of measurements in man, also reflect the uncertainty of animal experimentation and realization that the study of man alone can ever prove entirely valid for other men. [77]

In the same book, a British colleague, Sir Douglas Black, agreed, "I share Dr. Dayan's skepticism about the value of much of the animal toxicology we do now in predicting toxicity in man." [78]

Practically any scientist who is familiar with the disadvantages of animal testing would be obliged to admit the truth of the following words:

> Unfortunately, there is no absolute certainty that a substance that has not been found to produce any adverse effects in animal models will also be safe for humans. . . . The use of animal testing is often criticized on several levels simultaneously. Examples have been frequently presented in the literature as to how medically important drugs were discovered by luck in humans that would not have been found by current testing procedures had the initial studies been with animal models. . . . There are inherent problems in trying to make interspecies comparisons. By taxonomic definition, a rat is a rat and not a human. Human beings have been reproductively isolated for millions of years, and in addition to reproductive differences, numerous other metabolic differences have also developed; in fact, these differences may be so significant as to cause rats and humans to respond very differently to a same agent. [79]

The drug-testing process is far from flawless aside from the animal-testing phase. Though those egregious inadequacies are beyond the scope of this book, please note that physicians have long criticized the drug industry for abbreviating the clinical trial stage. This is where many severe side effects reveal themselves and is also the most reliable but expensive stage. By way of explanation, the General Accounting Office report stated that of ten prescription drugs withdrawn from the market between 1997 and 2001, eight injured women at a rate greater than men. Raxar and Duract affected men and women equally, but Lotronex, Redux, Rezulin, Pondimin, Posicor, Seldane, Hismanal, and Propulsid were more detrimental to women. This confirms that more women should be included in clinical trials, as was mandated in 1992 but is not happening. "This finding shows serious and unacceptable cracks in the FDA approval process," said Senator Tom Harkin. "The agency clearly needs to do a better job of determining how drugs affect men and women differently. For some women, this can be a matter of life and death."[80] (Women are not included in clinical trials, in part, because the manufacturer is concerned that the woman may, unknowingly, be pregnant and the test drug may damage her unborn baby, thus exposing the manufacturer to liability.)

There are additional difficulties. A recent article in the journal of the American Medical Association (*JAMA*) revealed that physicians involved in clinical trials regularly under-report drug safety problems. The authors of the study states, "We found no instances where the safety reporting can be deemed satisfactory." Physicians adequately report side-effect severity only thirty-nine percent of the time and toxicity only twenty-nine percent of the time. Only forty-six percent of the time do they say why patients withdrew from the trial.[81]

Why the Animal Model Fails

Why do animal systems fail to predict what a drug will do when given to a human? The reasons are plentiful. Animals differ from humans and from each other in respect to their anatomy and physiology on the cellular, sub-cellular, and receptor levels. Their habits are inherently different; they may be nocturnal or diurnal, and they may flourish on foods and portions that are very dissimilar to those humans require. These differences result in non-corresponding absorption, distribution and metabolism of substances. Conditions in labs are highly controlled, more so than in human life. And the dosages researchers administer to animals may be much larger than human dosages, by respect to body weight. Or, the human dose may be too high to give to animals.[82] Once again, we see that small differences on the cellular level, predicted by the theory of evolution, invalidate extrapolation between species. J. Caldwell stated:

> It has been obvious for some time that there is generally no evolutionary basis behind the particular-metabolizing ability of a particular species. Indeed, among rodents and primates, zoologically closely related species exhibit markedly different patterns of metabolism.[83]

The very methods that make lab animals so convenient and inexpensive are often part of the problem too. Here's one example: Lab mice are bred, through hormone dosages, to produce large litters. This makes them economical to cultivate and increases the profit margin from sales. However, the process appears to leave the mice tolerant of some environmental chemicals, specifically environmental estrogens or "endocrine disrupters," which are problematic for humans. Tests using these prolific mice fail to disrupt function in the mice because, some scientists feel, their breeding has rendered them resistant to estrogens.[84] Additionally, metabolism speed is a variable. Lab animals are smaller than humans and must metabolize faster to maintain body temperature stability. As a result, they eliminate some toxins more rapidly than humans, often before the toxicity occurs.

We have described how some genes govern the production of enzymes and how drugs are created around certain drug-metabolizing enzymes. Also touched on was the fact that enzymes are distinct, in quality and quantity, to each species. With versions of similar enzymes differing from one animal to another, and versions that change depending on what chemical is introduced or subtracted, the predictability of animal models can and does become even more abstract. In a book on extrapolation between species, the author wrote:

> An assessment of interspecies variability in the metabolism of foreign compounds must focus on the qualitative and quantitative differences in the activity and specificity of the enzymes catalyzing both phase I and phase II reactions. As Williams aptly pointed out, such interspecies differences in the metabolism of foreign compounds may result from 'differences a) in the absolute amount of an enzyme, b) in the amount of a natural inhibitor of the enzyme, c) in the activity of an enzyme reversing the reaction, and d) in the extent of competing reactions using the same substrate.'[85]

For instance, there are more than seventy known varieties of the important DMEs, liver enzymes called P-450 enzymes (which we discussed earlier) in different species. "Chicken-or-the-egg" style, the enzyme variations create physiological discrepancies and the physiological discrepancies create enzyme variations.

Species differences pose other impediments:

- The compound's solubility may limit its absorption in a smaller system.
- Pre-saturated enzymes may decrease absorption.
- Michaelis-Menten kinetics may be applicable and the blood levels may be greater than predicted in animals.
- Metabolites formed in the animal studies may cause toxicity that would not occur with the lower doses prescribed to humans. When an animal is tested, it usually receives a higher dose per body weight than a human does. This relatively higher dose may produce larger amounts of toxins that would do no harm in humans. (High doses of phenacetin produce this effect, among many examples.)
- Detoxification mechanisms in the liver or elsewhere may be depleted or saturated. (Examples are high doses of acetaminophen in the liver or of hexavalent chromium in the lungs.)
- Bioavailability of the dose form may be entirely disparate at lower doses due to local physiological effects (such as irritation) in the high-dose animal studies.
- High doses in animals may overwhelm organ systems, which would not be affected at lower doses, causing effects that mask those seen at lower blood levels.

Like other laboratory animals, rats share many physiological and anatomical characteristics with humans. However, there are some anatomical or physiological peculiarities that affect compound absorption, pharmacokinetics and metabolism, or cause unexpected reactions to a test compound. Rats are obligate nose breathers, for one; this can alter how and when a substance enters the blood stream. The placenta is considerably more porous in the rat than in the human. Owing to differences in the distribution of intestinal microflora within the rat gut, they are much more likely to metabolize an orally administered compound, possibly into a toxic or active metabolite. The rat's stomach acid secretion is continuous, whereas the human stomach secretes acid only in response to food or other stimuli. Also, rats are nocturnal, prone to different diseases than humans, have different nutritional requirements, and cannot vomit. David E. Semler of G. D. Searle and Company, writing about these discrepancies in *Animal Models of Toxicology*, goes onto describe differences between male and female rats, between different strains of rats, and between the results of studies on rats conducted at different research institutions. Even when the same rats are used for the same experiments at different research institutions, the results are different.[86] This is really no surprise to any informed person, as the book on extrapolation points out:

> That species may differ in their abilities to absorb compounds via the digestive tract should not be unexpected, since the gastrointestinal tracts of various species differ anatomically, physiologically, biochemically, in type and quantity of gut flora, and in the occurrence of different bile rates and acids.[87]

A drug that is rapidly absorbed in the GI tract of the rat may be poorly absorbed by humans or vice versa. The book examines the absorption of thirty-eight drugs and concludes:

> These examples should serve to illustrate that interspecies differences in the absorption of organic compounds do indeed exist and that in specific instances the degree of difference can be appreciable. Of the thirty-eight organic compounds reviewed here, more than one third of them . . . appeared to be differently absorbed by animal models as compared to the human subjects. To what extent organic environmental contaminants are differentially absorbed by animal models and humans and how such differences affect toxicity and carcinogenicity outcomes has not yet even begun to be aggressively researched. The experience in the drug field, though, does illustrate that species may differ substantially in their capacities to absorb organic compounds, and it is only logical to predict that such interspecies differences will be operational in the field of environmental toxicology of organics as well.[88]

It continues:

> In light of the striking influence that gut flora may have on toxicological outcomes, it is important to evaluate to what extent commonly used animal models differ with respect to humans in the type, location, and relative numbers of gut flora . . . it has been shown that species may markedly differ with respect to the type, number, and location of microflora. It has also been amply demonstrated that microflora may actively metabolize numerous compounds in toxicologically significant ways. . . . The extent to which metabolism of xenobiotics may differ between species is dependent on the number of microbiotic species, their location in the alimentary tract, and the specific substrate considered, to mention but a few variables. While quantification of interspecies differences may be difficult, an overwhelming amount of evidence indicates that the differential metabolism of xenobiotics by microflora may affect the occurrence and extent of enterohepatic circulation, the biological half-life of substances, and the promotion of toxic and carcinogenic substances.[89]

Apropos of placentas, one of the important thresholds drug developers need to cross is that of teratogenicity testing, meaning an exploration of what happens to the fetus when the mother receives the drug. Unborn children are highly susceptible to toxicity, usually more so than their mothers. Any substance, administered in sufficient quantity will damage the unborn child. Teratogenicity testing is very difficult. Had an effective means of conducting teratogenicity testing been available, the thalidomide disaster would not have occurred. But there was not. Think again of thalidomide's horrendous toll.

Animal modeled teratogenicity testing fails because there is very little parity between species on the level where drugs work. The National Academy of Sciences stated, ". . . no presently used species, including simians, resembles man in all of these respects. . . . It is evident that the degree of similarity to man exhibited by a given species varies from one test substance to another."[90] Since researchers cannot determine with certainty which animal model will offer the most accurate predictions for humans for any given substance, they usually select species for practical considerations, like cost and availability. Ergo, the dependence on rats and rabbits. One reason the results of teratogenicity testing vary between humans and rats and rabbits, the most commonly used species, is that rats and rabbits have very different placentas from humans.[91] Rats did not predict thalidomide's effects, as we have said, and only one obscure rabbit sub-species did, after receiving dosages far in excess of those that would have been given prior to sending the substance to clinical trials.[92,93] *Principles of Animal Extrapolation,* a book on animal models attempts to circumscribe the challenge:

> Teratological evaluation is a highly complex process because the potential mechanisms by which a substance may cause an adverse effect are exceptionally diverse. Wilson has listed thirteen stages that constitute the reproductive processes of mammals. . . . According to Wilson, the selection of the animal model must consider all aspects of the reproductive process if the findings are going to be applied to humans.[94]

Obviously, this is impossible. These facts confirm the impossibility of testing human teratogenicity through nonhumans. However, *in vitro* testing can give us a good idea of whether a drug will be teratogenic or not.

The relatively low percentage of drug-induced birth defects today is not due to pre-marketing tests as much as it is the realization by the public and the medical community that pregnant women should take only drugs that are necessary for life. More physicians now know that any drug has the potential to cause a birth defect; so they discourage expectant mothers from gambling with their infant's well being.

Testing for nerve tissue damage is likewise challenging in both live animals and in humans, but for different reasons. First, the blood-brain barrier prevents extraneous matter from entering nerve cells. Second, chemicals that are toxic to human neural tissue do not necessarily affect other species the same way. For example, the endoneurium of mice and rats excludes certain protein markers after intravenous injection while the same barrier in hens, hamsters, rabbits, cats, monkeys, and chimpanzees does not.[95] Many compounds, widely used in agriculture and industry, affect cells negatively, in addition to their primary effects on enzyme systems. Changes occur in both the central and peripheral nervous systems, but there is considerable variation among different animal species.

When opioids are administered to humans, primates, rats, rabbits and dogs, a panoply of effects occurs—central nervous system (CNS) depression, respiratory depression, constricted pupils, hypothermia, and a lower heart rate among others. In contrast, when opioids are administered to cats, horses, ruminants and swine, they stimulate the nervous system. The animal pants, its heart rate increases, and its pupils dilate.[96] Valium can cause liver failure in cats, but does not frequently do so in humans. The addictive properties of benzodiazepines, Quaaludes, and the opioid Talwin were not predicted in animal tests. Veterinarians Gordon J. Benson and John C. Thurmon conclude:

> Considerable variation occurs among animal species in their response to drugs used to alleviate pain and distress. The comparative pharmacodynamics and kinetics of most agents are unknown for many species, especially the smaller laboratory animals. Extrapolation of data from one species to another is fraught with error and should be avoided.[97]

Cats cannot take acetaminophen because of a relative deficiency in their glucuronide-conjugating system, obviously a deficiency not shared by the millions of humans who use acetaminophen regularly. And consider the metabolism rate for common aspirin: Aspirin has a plasma half-life of one hour in the horse, eight hours in the dog and thirty-eight hours in the cat. These variations are extreme. Aspirin never would have made it out of the box had animal testing been a requirement one hundred years ago when aspirin was discovered. It is for this reason that Op Flint of Bristol-Meyers Squibb Pharmaceutical stated in *Neurotoxicology In vitro,* "... it is impossible to establish the reliability of animal data until humans are exposed."[98]

Aside from teratogenicity, nerve tissue damage and a few other venues, some animals will give the same results to drugs as humans. And granted, animals and humans have things in common, so this should come as no surprise. The question is, *how predictable are the animal models?* The former director of a British research facility, Huntingon

Research Centre (now Huntingdon Life Sciences), commented, ". . . the best guess for the correlation of adverse reactions in man and animal toxicity data is somewhere between five and twenty-five percent."[99,100] There is that "twenty-five percent" again. As we state throughout this book, in order to be useful, a model must have a high degree of predictability. Twenty-five percent is not a high degree. The animal-model convention, in addition to being an inordinate stretch, is an anachronism. It is as though we still insisted on sending canaries into mines to test air quality. Yes, canaries are more sensitive to carbon monoxide (CO) than humans. However, continuing to use animals for pharmacokinetics is like using a canary instead of a mass spectrometer to detect CO.

Alternatives to Animal-Modeled Testing

Today, we have amazing technology that can re-create pharmacokinetics on the cellular, sub-cellular and even genetic level. Scientists' ever-more comprehensive view of receptor physiology, the structure-activity relationship, and physicochemical activity is paving very effective inroads to drug testing. These vehicles are as follows:

Studying bacteria—Presently scientists can identify different drugs harming different parts of the body, but they are not yet sure (in most cases) as to the mechanism. The underlying means by which the harm occurs may be through damage to DNA. This is called genotoxicity. The Ames test has become the classic test for genotoxicity. Its second generation, the Ames II test developed by Xenometrix is an improvement on the original. Researchers fill 384 wells on a plate with DNA and watch for mutation via pH change while a computer automatically analyzes the data. The samples run faster and with more predictive results than animal tests. Both Ames and Ames II use the bacterium *Salmonella typhimurium*. Another test using bacteria is Vitotox by Labsystems. It is able to measure both genotoxicity and cytotoxicity. Vitotox's advantage over the Ames test is that it targets all the DNA in a cell, whereas Ames is limited to genes related to the synthesis of a single amino acid, histidine.[101]

Computer-generated analysis and mathematical modeling—Scientists simulate the likely impact of a substance on a human in great detail and with astounding accuracy. Pfizer weighed the accuracy of a series of literature-based and in-house mathematical models using data from eighty-three Pfizer drug candidates over the past ten years. For example, the team calculated the apparent total body clearance value of the drug candidates—an indicator of how much the drug will be metabolized in the liver—from *in vitro* measurements using human hepatic microsomes (liver cells) and, in a few cases, animal models. They then used twelve mathematical models to predict human *in vivo* clearance. The most successful runs were accurate in up to eighty-eight percent of the 83 drug candidates, the computer-generated model predicting a value within twofold of the value measured in humans. Computer models of other characteristics, including volumes of distribution, half-life, and bioavailability, were also highly accurate.[102]

Physiochemical analysis of the prospective drug—Scientists predict the toxicity of a compound by analyzing its pH, partition coefficients, absorption spectra and other physical data. For example, angiotensin-converting enzyme (ACE) inhibitors are now an essential hypertension treatment. With few side effects, they block the formation of a chemical that constricts blood vessels. The scientists who made the ACE inhibitors possible, David W. Cushman and Miguel A. Odentti, wrote:

The active site model that we described in our original studies used simple chemical concepts guided by a hypothetical 'paper and pencil' model of substrate and inhibitor binding to enzyme. This rational design approach led to a class of structurally simple compounds that can inhibit the action of the enzyme with great potency and specificity, properties that translate *in vivo* into effective antihypertensive activity with a remarkably low level of unwanted side effects or toxicity . . . clinical research and drug discovery have always been closely associated.[103]

Cell viability testing—Scientists measure the effect of the test chemical on DNA and RNA synthesis, membrane integrity, cell division, and cell metabolism. The Neutral Red Release assay, the Lowry method, Coomassie Blue, Kenacid Blue and the MTT assay are all examples of *in vitro* tests used to assess cell viability. They can examine the effect of a drug on specific receptors or enzymes in human cell cultures. As P-450 enzymes, DMEs are involved with many drugs, and they are a prized commodity for drug testing, as we have mentioned. Liver cells usually lose their ability to metabolize drugs after a short period of time, however they can be genetically modified to produce specific cytochrome P-450 enzymes indefinitely.[104] A ten-year multi-center study involving twenty-nine laboratories in fifteen countries proved that cell culture tests are more accurate in predicting human response to drugs than are animal models. Also these *in vitro* analyses of cell viability are direct and simple compared with *in vivo* models.[105] Or as Mark R. Cookson of the University of Newcastle, stated:

One of the great advantages of *in vitro*, as compared to whole animal studies, is the relative ease by which biochemical changes can be measured and manipulated. Cell cultures lend themselves to detailed biochemical dissection, especially where the actions of a toxicant on one or a few particular types of cell is needed.[106]

Subcellular activity analysis—Scientists look for changes in target macromolecules and analyze biochemical responses. EYTEX and SKINTEX are examples of this genre of test.

The biotech creation called the DNA microarray, which we discussed earlier in the chapter, also shows what toxins can do to our genes. Though toxicogenomics, as it is called, is not yet sophisticated nor affordable enough to use broadly, studying genes thus could become a faster, cheaper, and more accurate way to test drugs, chemicals, food additives, and cosmetics. The DNA chip also allows scientists to apprehend how cell components respond to the drug in concert.

Patterns of gene activity displayed on the chip should, at least in theory, indicate whether the chemical is toxic or not, much as DNA fingerprints are used to judge the guilt or innocence of criminal suspects. Scientists can observe whether a drug turns on a harmful gene that then, say, starts uncontrolled cell replication, or whether it turns on a gene that kills cancerous cells. They can appreciate whether the drug is efficacious or injurious.

The microarray technology still needs refinement before it can become an everyday phenomenon. Genes constantly turn on and off as the body carries out its work. Some gene changes may indeed indicate the cell is in its death throes. But other gene changes could be part of a response that neutralizes the chemical so it does not cause harm. To form more definitive conclusions about the patterns, scientists are testing hundreds of chemicals with known toxicity, developing a database of genetic signatures against which the unknown compounds can be compared. Many start-up biotech firms are involved in this process—Affymetrix of Santa Clara, Tularik, Synteni and Incyte of Palo Alto, Hyseq of Sunnyvale, and the federal government via the National Human Genome Research Institute.[107,108]

Toxicogenomics is still in the early experimental stage, but it could render many advantages over current approaches. Present tests, for instance, may be able to determine that a substance causes liver damage, cancer, heart problems or birth defects, but they do not tell why it does so. The pattern of gene activity offers clues to the biochemical pathways by which the harm occurs. The detail is impressive. Microarrays quickly distinguish between two types of toxicity that would normally require time-consuming electron microscope examination of cells. Importantly, changes in gene activity may occur well before other more visible symptoms of harm, like tumors, which can take months to develop. So these tests would pick up toxicity other tests fail to spot. Also, gene tests are more sensitive to lower doses than other tests.

Though we must wait for toxicogenomics to be perfected, existing alternatives to vivisection make the animal model's inefficiency and dangers, by any measure, flagrantly obsolete. Consider the incidence of nerve tissue investigation, so impossible through animal models, as discussed. "Toxicologists have developed a large panel of methods to characterize neuronal death *in vitro*," wrote several scientists in *Neurotoxicology in Vitro*.[109] These have unique advantages. Scientists can examine Schwann cells and oligodendrocytes *in vitro*, wherein both retain their basic physiological properties and functional capabilities. Scientists culture nerve tissue, such as that obtained from dorsal root ganglions, then expose it to the medication being analyzed. The cells can be studied in conjunction with intact axons or axonal plasma membrane preparations from different nerve cells or with culture media composition (including growth factors and mediator substances) controlled to induce proliferating or differentiated states. These provide a range of possibilities for detailed mechanistic neurotoxicological investigation.[110] Either way, scientists can observe drugs' action directly on nerve tissue.[111] *In vitro* methods also bypass the interspecies differences in the barriers that exist around neural tissue, the varied blood-brain barrier just mentioned. It just makes a lot more sense than the animal model, as the following quote further elucidates:

> Because of the increasingly observed species differences in metabolism [during the development of the drug testing process], in pharmacokinetics and in mechanisms of toxicity, a variety of animals had to be exposed, including rodents, rabbits, dogs, monkeys and baboons, to warrant that no aspect of toxicity was missed. . . . Yet this extensive screening in various experimental animals for potential toxicity in humans still threw up numerous false negatives, such as the drugs perhexiline, benoxaprofen and tienilic acid, which subsequently caused human deaths. . . . So, there are indeed more appropriate alternatives to experimental animal studies and, for the safety evaluation of new drugs, these comprise short-term *in vitro* tests with micro-organisms, cells and tissues, followed by sophisticated pharmacokinetic studies in human volunteers and patients.[112]

Why Animal-Modeled Testing and Drug Development Continues

Clearly, the technology and know-how to screen drugs without animals exists. Present *in vitro* tests and cellular models are, by rapport to the animal model, very accurate.[113] And better *in vitro* tests are on the horizon. The technology—extant, reasonable, and far superior—is thwarted by myriad obstacles, some anticipated, and some simply put there to preserve the protocol for animal testing.[114] Moreover, animal testing remains mandatory in some instances. The ever-antediluvian FDA still requires animal-modeled teratogenicity and carcinogenicity tests.

It is not just inertia that keeps the FDA prehistoric. Clearly in some cases, the agency appears to work more for the pharmaceutical industry than for the American public. And the pharmaceutical industry refuses to give up massive profits for increased patient safety. In a noted book on the risk-benefit analysis of drug research, an executive at Smith, Kline and French was quoted as saying:

> What I am trying to get at is a situation where we do not automatically accept the traditional toxicology. . . . I think I am saying that the present tests are well known to us but that does not make them good. There may be better tests around, but we have no incentive whatsoever to look for them at the moment. In fact, quite the reverse [because of the fear that regulatory agencies will require the new tests in addition to the current ones].[115]

That is to say that in order to screen drugs pre-clinically more *in vitro* tests would have to be done. It would take longer and cost more. Again, the only advantage would be patient safety, not necessarily a priority that supercedes profitability. If more rigorous pre-clinical tests were mandatory, it would tighten the market. Knowing that more expense would be involved, pharmaceutical companies would have to think twice before propagating "me too" drugs. We do not need more of the same drugs, such as nonsteroidal anti-inflammatory agents that work only as well as the other ten or twenty we currently have. Certainly, that helps "Acme Drug Company" get their share of the nonsteroidal anti-inflammatory market, and thus makes shareholders happy; but it exposes more people to risk of a new drug and medically benefits no one. Yet, without regulation to the contrary, pharmaceutical companies and their staff, as well as research institutions and their staff, stubbornly cling to the old style. As clinical toxicologist Dr. Roy Goulding put it:

> Today the subject and practice of toxicology have become exalted to the eminence and influence of a religion. It is, moreover, an established form of worship, actively supported by the State. It has its creeds and its commandments, and its hierarchy of high-priests, worshipers, adherents and novitiates. Again, like a religion, it relies more on faith, than reason.[116]

Excessive relabeling and drug-recall persist because the approval process remains biased by the pharmaceutical companies that pay for it.

One proof of this may be found in an examination of cost and cost-effectiveness reports for oncology drugs published between 1988 and 1998. It found significant differences in conclusions between the studies paid for by pharmaceutical companies and those sponsored by nonprofit organizations. Only five percent of corporate-sponsored studies reached unfavorable qualitative conclusions, compared with thirty-eight percent of nonprofit studies.[117] *He who pays the piper calls the tune.* Not surprisingly but still appallingly, drugs, whether questionable or not, stand a far greater chance of reaching the market—seven times the chance—when pharmaceutical companies pay for their testing. And a primary mechanism drug companies use to support the merit of their product is animal testing. The researchers also found that corporate-sponsored studies were more likely than nonprofit-sponsored studies to overstate quantitative conclusions. In other words to draw ". . . a favorable qualitative conclusion when quantitative results were neutral or unfavorable, or a neutral conclusion when quantitative results were unfavorable."[118] Biases are much the same in the United Kingdom. The *Times* reported in the UK on August 8, 2000, that two-thirds of the members of the government's pharmaceuticals-licensing board have financial interests in pharmaceutical companies. If such allegiances, which potentially compromise consumers, exist in the light of day, think how easy it is to falsify data, or do research that has no hope of ever benefiting society behind the closed doors of an animal lab.

Sacred Cows and Golden Geese explains that research facilities are disposed to supply pharmaceutical companies and other industries with favorable results because they rely on them for research dollars. Telling a corporation that their product is substandard might obligate the corporation to seek another research facility for the next product. So research facilities try to supply positive data. Frequently, the compatibility extends even further. Ten of the U.S. medical schools that receive large amounts of research funding from the NIH are the University of California at San Francisco, Baylor College of Medicine, Columbia University College of Physicians and Surgeons, Harvard Medical School, Johns Hopkins University School of Medicine, University of Pennsylvania School of Medicine, University of California at Los Angeles School of Medicine, University of Washington School of Medicine, University of Washington School of Medicine at St. Louis, and Yale University School of Medicine. Of these, only one prohibits its research faculty from holding stock and stock options in sponsoring companies, though the reference did not distinguish which.

Dr. Jeffrey M. Drazen, editor-in-chief of the *New England Journal of Medicine,* which published these results, noted that the differences in policies governing financial arrangements between sponsors of clinical research and investigators "create the potential for an uneven playing field among academic medical centers." He continued, "The lack of strong, consistent policies could erode the public's trust in those studies, profoundly affecting the future of clinical and basic research."[119]

Sometimes institution regulations on researcher investments are soft. For instance, the institution might generally ban financial interest in a company developing a product being studied but permit exceptions "for newly recruited faculty members and for family gifts or inheritances." And only a minority of research centers requires initial disclosure to research sponsors, funding agencies or journals. Needless to say, this results in pervasive conflict-of-interest.[120]

And that is just the academic centers. The last decade has spawned a whole new type of collusion between commerce and research—for-profit companies and research institutes run by the same people. New non-profit research centers are proliferating, often right next door to the for-profit corporations that benefit from their research, as pointed out in an article in the *Wall Street Journal*. Recognizing the "strong tradition of government research money generally going to academic and non-profit institutions," entrepreneurial scientists have learned how to manage both nonprofit and commercial corporations:

> [They] have become adroit at using private nonprofits to obtain government grants and then using the money to fund research that helps their own for-profit biotechnology ventures. In case after case, scientists closely linked to profit-making companies are applying in the name of nonprofit institutes to obtain multimillion-dollar NIH (National Institutes of Health) grants.[121]

Chairpersons at these nonprofits and for-profits claim there is no conflict of interest, but, as is often the case, the same people sit on both boards or chair the board of one while working at another, their claims tend to breed suspicion. Creating positive results to push products into the marketplace is the first priority. If businesses can use taxpayer dollars to pay for those results, why not? And barring more exigent demands from the FDA, they can get away with animal-modeled results to hasten their innovations through the approval process.

Whereas for-profit companies need to heed their financial bottom line, nonprofit research institutions do not have to be as cautious, especially as NIH funds are expanding. The NIH is our government's greatest reservoir of medical research monies. NIH awards to commercially linked nonprofits rose over fifty percent between 1994 and 1999. All signs indicate they will continue to do so. NIH has fifty-seven percent more money this year than it did five years ago, and Congress is pushing to augment its annual budget by another fifty percent by 2003, to about $27 billion.[122]

Following a controversy in which Sandoz Pharmaceuticals agreed to pay $300 million over a period of ten years to Scripps Research Institute in exchange for first rights to license most of the institute's biomedical research, the NIH issued guidelines for nonprofit grants in 1994. The guidelines say that nonprofit recipients should not give commercial companies the ability to direct their research agendas. However, as Dr. Ruth Kirschstein, the NIH's acting director pointed out, the NIH does not actively enforce these guidelines.[123] So the inevitable happens. As is apparent in examples throughout this chapter, it is not hyperbole to suggest that the FDA effectively colludes with pharmaceutical companies to bolster consumer confidence in certain drugs where no confidence is really due. And the convention they use to instill confidence is animal experimentation.

Another syndrome that keeps animal testing labs and their suppliers in business is the media. Medical "miracles" grab virtually everyone. Hence, Television and newspaper reports of prospective new drugs exaggerate their efficacy and minimize the obstacles and side-effects.[124,125] Considering the slanted press releases the editors receive, it is easy enough to do. These news items almost invariably involve animals. Apparent advances garnered through animal studies get lots of publicity, but when these same drugs later perform poorly in clinical trials, there is barely a whisper. It is no wonder, since drug companies certainly do not generate press releases advertising their failures. Adverse reactions to prospective treatments and safety information, such as potentially dangerous side effects, get, on an average, about a third of a page in medical journals. A study conducted by Dr. John P. A. Ioannidis of Tufts University School of Medicine and Dr. Joseph Lau of the New England Medical Center in Boston examined a total of 192 randomized drug trials, each involving a minimum of a hundred patients with at least fifty patients per study arm. Just thirty-nine percent of randomized drug trials adequately reported clinical adverse effects of the medications being studied. Only twenty-nine percent did a good job of reporting the toxicity of a drug revealed through an abnormal lab test result. Eleven percent and eight percent only partially reported clinical adverse events and laboratory-determined toxicity, respectively, the team noted.[126]

Medication testing on animals continues for the same reasons that any failed politics-based agenda continue. Positive animal-modeled results work short-term by inspiring public confidence. This allows pharmaceutical companies to sell their products, while effective public relations obscure the negative long-term consequences of using those products. Animal experimentation provides jobs, as well as research money for universities and other testing facilities. It gives liability protection to the manufacturers. It benefits opportunistic politicians who are eager to appear concerned about America's drug supply. And it gives long-established bureaucracies the appearance of legitimacy.

chapter 5

Surgery

I agree that for the benefit of medical science, vivisection or animal experimentation has to be stopped. There are lots of reasons for that. The most important is that it's simply misleading, and both the past and the present testify to that.[1]

—Moneim Fadali, M.D. (cardiac surgeon)

If animals were no longer available for testing, would the development of new surgical procedures cease? Of course not. Yet, the misconception that feats of scalpel and forceps must firstly occur through fur prevails. The idea is predicated on uniformity between species that simply does not exist.

As we have noted before, organs are arranged in more or less the same configuration throughout the Family *Mammalia*. And all mammals have many of the same properties, especially those visible to the naked eye. However, we now know that the interior of individual species is also fraught with idiosyncrasies, both visible and microscopic. These distinctions make the translation of techniques from animals to humans a clumsy enterprise, filled with guesswork. Procedures that do not work on animal models can be discarded, only to be performed successfully later in last-ditch efforts to save a human. On the other hand, operations that well-intended surgeons perfect on animals sometimes prove lethal to humans.

This chapter will share a better understanding of the real methods that have produced viable, marvelous surgeries throughout time. These facts themselves will dispel the belief that practicing on animals is compulsory, because in reality, surgical innovations originated on humans by necessity and did not require animals.

Early Surgery

Experimentation on man is usually an indispensable step in the discovery of new therapeutic procedures or drugs. . . . The first surgeons who operated on the lungs, the heart, the brain were by necessity experimenting on man, since knowledge deriving from animal experimentation is never entirely applicable to the human species.[2]

Looking back over the history of surgery, we find it has always been as thus described by René Dubos, Pulitzer Prize winner and professor of microbiology at the Rockefeller Institute of New York. Surgical procedures were most often forged on patients whose dire condition did not allow time for trial and error on animals.

Reading texts that describe surgical innovation, one cannot help but be impressed with how many original surgeries were performed on a battlefield or in other life-or-death situations. Ancient times being violent, as they were, the first surgeries were probably amputations. Hence, the dilemma of severed vessels provided the earliest surgical challenge. Historians credit a seventh century B.C.E. Indian surgeon named Sushruta as the first to tie off blood vessels so they would not bleed during surgery. Progress in the West was slower. Not until the mid-1500s did Ambroise Paré demonstrate that it was not necessary to treat wounds with boiling oil or cauterize them with a red-hot iron to control bleeding. Paré developed a technique based on clinical observation. With a few simple knots, he tied off bleeding vessels, as his Asian predecessor had done over two thousand years before. Records exist of a man named Assman of Kroningen who attempted to repair femoral artery lacerations in dogs in 1773.[3] All the animals consistently thrombosed, according to the literature, thus the procedure never gained favor. In 1803 David Flemming, a naval surgeon, performed the same operation, tying off a lacerated carotid artery on a human patient who was bleeding to death.[4] Unlike Assman of Kroningen, his patient did well. Nevertheless, the animal-model community still plaudits Sir Astley Cooper with ligating (suturing closed) the first carotid artery of a human in 1905, based on animal experiments he had performed previously.[5]

Most diseases were unexplained and untreatable in previous centuries. Physicians rarely got out the scalpel, except on wounded soldiers or on the catastrophically ill, whose conditions were also of a critical nature. Ephraim McDowell was one of the more intrepid; he performed the first oophorectomy (removal of an ovary) in 1809 in Danville, Kentucky. Though McDowell published his accomplishments in a medical journal, he was ridiculed because no one believed they were possible. McDowell is now considered the father of abdominal surgery.[6] His triumph launched an epoch of surgical innovation.

Surgery had been kept in stalemate by patients' inability to withstand pain and by mortality brought on by infection. Outside rote amputations, surgical procedures did not measurably advance for hundreds of years simply because people could not tolerate the excruciating agony or because they succumbed to sepsis (infection). Nineteenth century developments would overcome these hurdles with sterile technique and anesthesia, both immense breakthroughs. With the coming of the twentieth century, modern technology—most specifically in the field of radiology—increasingly allayed the element of chance in surgery.

Innovation in Sterile Technique

Asian and European medical practitioners had realized, far before medieval times, the importance of keeping the sick area clean. Greek physicians followed Hippocrates' example, scrubbing both their hands and the infirmary. However, following the fall of the Roman Empire, these precautions lapsed inexplicably, much to the detriment of patients and others infected with their diseases as a consequence of bad hygiene. The battlefield and indeed hospitals themselves teemed with infectious opportunity:

> . . . the man laid on an operating room table in one of our surgical hospitals is exposed to more chances of death than was the English soldier on the field of Waterloo.[7]

Such was the horror of death from post-surgical infection.

Nevertheless, hygiene was a small matter for most surgeons since it was not part of their training. Though practices were, as always, resistant to change, several astute and brave clinicians were finally able to force alteration of the dangerous practices. An important influence was a Hungarian physician named Ignaz Semmelweiss who practiced in Vienna.

In those days, students and professors had finally come to realize the educational importance of autopsies. From the autopsy room they went directly to the hospital rooms and wards to tend their patients—without washing their hands or changing clothes. Although the risks of this nonchalance are obvious now, they were not then. People, Semmelweiss among them, did notice, however, that certain disease was most rampant in maternity wards. Although most women delivered at home, those who went to hospitals contracted puerperal infection, at the time a scourge throughout European maternity wards.

Semmelweiss remarked, in 1846, that patients attended by students had a much higher rate of puerperal infection than those in other areas. He also observed that the students' patients developed symptoms of blood poisoning. When a good friend and colleague who had inadvertently cut himself with an instrument from the autopsy room died of similar symptoms, Semmelweiss took action. He concluded that the students were ferrying infections from the autopsy room to the wards and mandated that they wash their hands prior to coming into the wards. His observations were so astute that the mortality rate in his hospital plummeted. Due to professional jealousy and resistance to change, Semmelweiss' supervisor dismissed him from the hospital staff and drove him from Vienna. [8,9,10] S. Peller states:

> In a world that had not been stultified by the idea that only animal experimentation and only the laboratory can provide proof in matters of human pathology, the battle against puerperal fever would not have needed to wait for the discovery of cocci (bacteria). The experts who, during the 1840s, opposed and prevented the initiation of a rational program of combating the disease, should have been charged with negligence that resulted in mass killings. But they were not. Instead they continued to enjoy the privileges and prestige previously accorded to them, while the benefactors suffered and paid dearly for their discoveries. [11]

In 1864, Louis Pasteur announced his germ theory of disease. His discovery of bacteria reinforced the logic behind Semmelweiss' actions. It provided the sound proof of invisible infectious organisms that would spur surgical practice toward sterility. Unfortunately, however, Pasteur's finding coincided with renewed enthusiasm for animal experimentation, due to Claude Bernard. (See *Sacred Cows and Golden Geese.*) Animal experiments of the time encumbered innovation in sterility, in two respects. First, because many animals are not as susceptible to infection as humans, results of animal experimentation were inaccurate; and second, animals cannot effectively communicate the nature and extent of their malaise. Therefore, any assessment of infirmity or therapy was partially a matter of conjecture.

In 1876, Joseph Lister crediting Semmelweiss' measures set about expunging the surgical area of bacteria, recently described by Pasteur. Lister started spraying carbolic acid, or phenol, onto the surgical field. Lister had tried this on animals, apparently with success. Drs. Botting and Morrison, representing the animal experimentation industry, wrote a prevaricative article for *Scientific American* on the subject in February 1997:

> . . . Lister, who pioneered the use of carbolic acid to sterilize surgical instruments, sutures, and wound dressings [using animals], thereby preventing the infection of wounds.

What Botting and Morrison did not mention is that carbolic acid turned out to be a terrible idea. Though, in theory, Lister's intent was sound, he had the wrong chemical. Phenol does kill germs, but it also kills living tissue. Further, since it was only applied to the wound area, living germs were still nearby. Phenols are still used in disinfecting cleansers today, including Lysol—certainly not a substance you would want dripped into any incision. Lister deserves credit for advancing the idea that certain procedures could prevent infection, but his animal model-derived results with phenol were inapplicable in humans. Though Lister's animal subjects could not complain of the ensuing numbness and pain caused by the phenol, the harsh chemical's toxicity worried many surgeons of the time. Instead of spraying phenol on the incision site, they reverted to the age-old technique of boiling the instruments, using clean clothing and towels, and draping the towels around the area to be operated on, as well as washing and rewashing their hands with harsh soaps. Many prominent surgeons, including Tait, Kellogg, Bantock, and eventually even Lister, agreed that this preparation known as sterile technique was more important than anything else. Once sterile technique was widely accepted, mortality decreased dramatically.[12,13,14,15,16]

Meanwhile, animal experimentation continually failed to reinforce the exigency of sterile technique. Animals simply did not get sick with the same frequency as humans after operations that exposed them to bacteria.[17,18,19] Although animals can certainly become septic, many species are amazingly tolerant of non-sterile surgical techniques. To know why, one has only to reflect on the sort of resistances species such as dogs that subsisted off bacteria-ridden food might have developed over the course of evolution. Animals with a resistance to infection would be highly selected over the course of time. But the lack of infection in filth-exposed lab animals reinforced the casual, non-sterile approach that characterized most surgery of the time and prevented the speedy conversion to sterile technique throughout the profession. It is still routine to operate in the abdomen of a cow in the barn. Surgery in a cow standing in a stanchion between her fellow cows rarely results in infection. Few humans could survive such treatment.

Nevertheless, a gradual shift to a cleaner surgical environment continued throughout the nineteenth century. Surgeons scrubbed assiduously with abrasive cleaning agents, many times a day. This practice was very hard on the skin and many physicians and their attending staff developed allergies. In 1889 a surgeon named William Halsted introduced the use of surgical gloves in his operating room because his assistant, whom he later married, suffered from a contact dermatitis. Recognizing that rubber gloves sterilize better than skin, Dr. Halsted contacted the Goodyear Rubber Co. and had them prepare rubber gloves for his use. As other surgeons adopted sterilized rubber gloves, the protection against germs increased. Thus, mortality and morbidity further declined.[20]

Development of Anesthesiology

Everyone knows that the term "surgeon's hands," rubber gloves aside, refer to fingers that can deftly operate. Before anesthesia was invented, though, people cared far less about their surgeon's dexterity than they did his speed, simply because they could not withstand the pain. Haste was the top priority, and surgeons were commended not for their finesse but for their velocity. Anesthetics were to revolutionize the world of surgery, making way for surgeons whose skills were meticulous and innovative, regardless of pace, and for complex procedures that required more time. However, like most aspects of contemporary medicine, the development of anesthesia was fitful and did not gain momentum until the mid-nineteenth century. Some of the delay was the fault of animal experimentation.

At one time, people relied on either heat or cold to alleviate suffering, techniques that we still employ today.[21] In the East, the analgesic properties of plants such as opium and hemp were familiar since a millennium before Christ. Poppy seeds, from which opium is derived, were sometimes effective but unpredictable. Likewise, it was very easy to overdose. In1806, Friedrich W. Serturner experimented with opium in order to purify it into a usable form. He self-administered the drug and found the form that best produced relief from pain.[22] In a few pages we will discuss how differently animals and humans respond to the opiate morphine.

One of the earliest documentations of anesthesia in Europe was in chickens. Those with a vested interest in animal models would have you believe the following distorted version of the events:

> The greatest single discovery leading to the alleviation of suffering and pain, that of anesthesia, was initiated, controlled, and perfected mainly by properly devised animal experiments.[23]

In actuality, the discovery was entirely coincidental, in no manner an intentional animal experiment. In 1540, Paracelsus used a substance known as *sweet oil of vitriol* to sweeten the feed of chickens.[24] Paracelsus noted that the chickens fell asleep after eating the new food but woke up unharmed. This substance, later named *ether*, would enable millions of surgeries but not for three hundred more years. The real story of the development of anesthesia is much different. And it is filled with all the jocularity that mind and body-numbing substances tend to induce.

General Anesthesia

It is hard to believe that many of our ancestors submitted to the knife after mere hypnosis, but it is true. Surgeons of the 1700s and early 1800s relied heavily on mesmerism as their not-altogether effective antidote for pain. But, earlier, in the 1770s, a chemist named Joseph Priestly laid the groundwork for modern anesthesiology by combining oxygen and nitrous oxide, thus inventing "laughing gas."[25] Soon after, an American chemist and physician, S. Latham Mitchill, administered the concoction to animals. So many creatures died that he determined the gas was poisonous and abandoned its development as an anesthetic. The fatality level even led Mitchill to believe that nitrous itself was a form of *contagion.*

In the early 1800s, however, a man named Humphrey Davy so doubted Mitchill's claims about nitrous oxide that he sniffed it himself— in ever-increasing amounts. Afterward, he wrote with enthusiasm about his giddiness and lack of conscious pain. Davy found particular relief from nitrous when he was cutting his wisdom teeth.[26,27,28] Davy's gusto and other people's self-experiments put laughing gas on the party circuit. It became a popular parlor drug in the early 1800s, provoking all kinds of hilarity. During this same time period many others were self-experimenting with nitrous and other gases and finding them harmless and entertaining.[29] But no one yet considered nitrous oxide as a serious aid for patients undergoing surgery.

Crawford Long, a general practitioner in Jefferson, Georgia, was also an early self-experimenter. He used nitrous oxide on himself and then used it to perform the first general anesthetic in 1842 while removing a tumor from a patient's neck. The surgery was painless. Long rushed to share news of this breakthrough, but influential surgeons who favored the hypnosis alternative kept his papers out of the medical journals. (Even then, the medical literature was not always objective and sometimes impeded change as it does now. There are also rumors that Long's discovery was kept out of influential journals because others learned of it and wanted to take credit for it themselves.) Finally in 1848, Long's article was published. Meanwhile in 1844, another physician, Horace Wells, used nitrous oxide on patients and on himself as well. A man named Gardner Colton administered the gas so Wells could have a tooth removed.[30,31,32,33]

Crawford Long also self-experimented with ether. His reasons for dabbling in volatile gases were not strictly Hippocratic. In the mid-1800s he wrote, "We have some girls in Jefferson who are anxious to see it [ether] taken," meaning they would like to take it themselves after watching others do so, "and nothing would afford me more pleasure than to take it in their presence and get a few sweet kisses."[34] After one of Long's first experiments on himself, he regained his senses with bruises present where none had been before. Long concluded that ether must be an anesthetic too.[35] Michael Faraday, a student of Davy's, was also among the first to describe ether's effect following self-experimentation.

Sometime prior to 1842 a dentist named Elijah Pope gave ether to a patient undergoing a tooth extraction.

In 1846, this covey of anesthesia experimenters found peer support. Horace Wells's partner, William T. G. Morton, administered ether at Massachusetts General Hospital in front of a distinguished crowd of observers. The surgeon conducting the operation is said to have exclaimed at the end of the pain-free operation, "Gentlemen, this is no humbug." This occasion is traditionally held to be the premier public demonstration of general anesthesia, although it had been used many times before, without such a prestigious audience. Following this watershed demonstration, Oliver Wendell Holmes suggested the actual word *anesthesia,* which means loss of feeling, to describe substances that brought on insensibility.

Other gases were concocted and quickly added to the list of inhalant general anesthetics. The discovery of the anesthetic properties of chloroform, as an example, came on the heels of human experience, documented thus:

> Late one evening—it was the fourth of November 1847—on returning home after a very weary day's work, Dr. Simpson, with his two friends and assistants, Drs. Keith and Duncan, sat down to their somewhat hazardous work in Dr. Simpson's dining room. Having inhaled several substances, but without much effect, it occurred to Dr. Simpson to try a ponderous material which he had formerly set aside on a lumber table, and which on account of its great weight, he had hitherto regarded as of no likelihood whatever. That happened to be a small bottle of chloroform. It was searched for and recovered from beneath a heap of waste paper. And with each tumbler newly charged, the inhalers resumed their vocation. Immediately an unwanted hilarity seized the party; they became bright eyed, very happy, and very loquacious—expatiating on the delicious aroma of the new fluid . . . a moment more and all was quiet—and then crash. The inhaling party slipped off their chairs and flopped onto the floor unconscious.[36,37]

Had Simpson rallied hounds around the table in lieu of colleagues, the evening and the history of surgery would have progressed entirely differently. Since chloroform poisons dogs, especially pups, it would have been withheld for many years from the services of man. Chloroform was soon put to use as a painkiller for childbirth.[38] An animal experimenter of the time named Flourens, in consequence of the fatal effects he observed in animals, discarded chloroform altogether as an anesthetic. Later, after chloroform was already in use on humans, leading English authorities ridiculed subsequent chloroform experiments on dogs. Their primary target was Sir Lauder Brunton, who suggested that chloroform did not work.[39]

One enormous drawback of early general anesthesia was that it impeded respiration. Unless anesthesiologists facilitated breathing, their patients risked brain damage and possibly death. Therefore, as anesthesia came into use, techniques for keeping air circulating through patients' lungs were required and conceived of by necessity. A certain surgeon needed to remove a tumor from the base of a patient's tongue so the anesthesiologist could not administer anesthesia in the conventional way with a mask. After first practicing on cadavers, he placed a tube into the trachea. This gave the surgeon access to the surgical area and kept the airway open. This, the *endotracheal tube*, as it was termed, also allowed the anesthesiologist to administer inhalation anesthesia through the tube directly into the lungs instead of with a mask or cotton gauze as had been done previously.[40]

When patients cannot breathe, such as occurs when muscle relaxants are administered during surgery, positive pressure ventilation is necessary. Animal experiments delayed the implementation of positive pressure ventilation for years because the procedure appeared too dangerous. Eventually common sense and human data won out.[41] World War I also provided more knowledge of positive pressure ventilation methods. The process was made easier by placing a cuff on the end of the endotracheal tube. It could be inflated, thus allowing pressure to be transmitted without interruption from outside the thorax to inside. Only after positive pressure ventilation was shown effective in humans did researchers "validate" the technique in animals.

Over the years, innovation, both in surgery and chemistry, resulted in new forms of anesthetics. Anesthesiologists weigh the extent of the new anesthetics' value during surgical procedure against whatever side effects they may produce, during and after the surgery. Change is an inevitable side effect of progress. For example, cyclopropane was the most widely used general anesthetic for over thirty years in the mid-twentieth century. It had been discovered accidentally when chemists witnessed its unique properties. But the use of cyclopropane was discontinued when electrical equipment started being used in the operating room because the gas causes explosions if exposed to a spark[42] It was replaced by better inhalation anesthetics such as halothane, ethrane, and isoflurane.

Local Anesthesia

By subduing pain, general anesthesia led to more abundant and inventive surgical procedures, as well as curiosity about other anesthetic measures. Soon after general anesthesia was put into practice, local anesthesia emerged, first in the form of cocaine. South Americans had long used coca leaves for pain relief. On a trip to that continent, a physician named Scherzer watched the native people chew the leaves as well as grind them up and apply the poultice directly to the skin. Scherzer brought large quantities of this miracle substance back home, thus spawning interest in the plant's numbing properties. Others then self-experimented and finally isolated coca's active agent, cocaine.[43]

In 1880, a physician named Vasili Konstantinovich Von Anrep published a paper describing cocaine's anesthetic properties.[44] Soon thereafter none other than Sigmund Freud demonstrated its effectiveness as a local anesthetic and suggested using it for patients undergoing surgery.[45] An ophthalmology intern named Karl Koller assisted Freud in his cocaine experiments. Koller placed the medication on his own tongue. Experiencing the numbness there, he deduced that cocaine would work well for eye surgery.[46] Koller also experimented on frogs, but the results only reproduced the earlier experiments that Freud and he had conducted on themselves. Drs. Richard Hall and William Halsted also injected themselves first before going on to use cocaine on their patients.[47]

Animals received a lot of cocaine in experiments, even then. Interestingly, no anesthetic property was apparent when the cocaine was applied to an animal eye, though experimenters did report dilation of the pupil. Cocaine worked very well in humans, however, and is still used as a local anesthetic. It is not as popular today because of the potential for abuse. This is unfortunate, because there are certain procedures for which cocaine remains the most effective local anesthetic.

Newly emerging local anesthetics continued to transform the practice of surgery during the late 1800s and into the twentieth century. As contrasted with general anesthesia, they allowed patients to remain conscious and to breathe on their own, thus avoiding some of the complications of general anesthesia. Lidocaine, still the most commonly used local anesthetic, was discovered by accident during this time. A pharmacist, who insisted on tasting all his medications (not uncommon many years ago), noticed when he first tried lidocaine, that his tongue became numb. The value as an anesthetic was apparent.[48] The pharmacist developed fifty-seven more medications similar in structure to lidocaine, based on chemical modeling. All local anesthetics used today are derivatives of lidocaine and cocaine. This is a common theme in medication development: Pharmacists locate a chemical structure that produces certain results, then perfect its effective properties through modifications.

Muscle Relaxants

Involuntary muscle movement can pose a significant surgical challenge, interfering with precision and accessibility. At first, anesthesiologists solved this problem by giving huge quantities of anesthesia to slacken all muscles. These quantities were dangerous because they slowed recovery from anesthetics and threatened the patients' respiratory system. Surgeons felt that muscle relaxants, which would allow better observation, would be enormously beneficial, particularly in abdominal operations such as bowel resections, appendectomies, or gall bladder surgery. And certainly muscle relaxants were absolutely essential to demanding operations involving the brain, eyes, or spinal cord, where the slightest twitch can be catastrophic.

As is often the case, researchers took their cues from primitive peoples, and again the jungle offered a prospect. Curare, one of the first muscle relaxants, was derived from a plant that was used in hunting by indigenous South American people. Though its attributes were evident in a 1740 description by Charles Marie de Lacondamine, animal experiments hampered its adaptation. One researcher learned that physicians used curare to facilitate the exams of mental patients in Nebraska (a practice considered very unethical today). He thought to give it to dogs and cats, all of which suffered extreme malaise, some to the point of death. However, they did not die instantly of respiratory arrest or paralysis of the muscles used to breathe, as humans who were not abetted with breathing devices would. These results prompted a Dr. Collier to say, "It was difficult to foresee from these experiments on animals how far a muscle relaxant was likely to affect respiration in man."[49]

Curare was also a puzzle because some thought it might be an anesthetic as well as a muscle relaxant. Under the effect of a muscle relaxant like curare, no one can communicate if or how much something hurts. Animals could not express whether curare numbed their pain or not. Dr. Frederick Prescott made the decisive move. He volunteered to take curare intravenously while he received ventilation so he would not die. Strips of adhesive tape were applied to his skin then ripped off, removing the hair beneath. After the curare wore off, Prescott stated emphatically that curare was not an anesthetic.[50] In 1942, Harold Griffith administered curare to humans and it worked in humans just as predicted two centuries prior.

One of today's most rapid-acting muscle relaxants, succinylcholine was almost overlooked due to animal experiments. When scientists first conducted tests to determine what effect succinylcholine had on the body, they tried it on animals that they had already paralyzed with curare. This masked the most important aspect of the new medication—succinylcholine too paralyzes. Once again, the use of animals as models for humans delayed release of an exceedingly useful medication, this time for more than forty years.[51]

Regional Anesthesia

Dr. August Bier, a nineteenth century physician, searched for a way to numb the surgical area as well as the surrounding region, but still leave the patient conscious. Such a technique would be more comprehensive than local anesthesia, and might provide advantages over full general anesthesia. Bier was convinced that injecting cocaine into the spinal column could anesthetize a larger area without rendering the patient unconscious. He tried this out on himself and the experiment was a success.[52] Unfortunately, Dr. Bier suffered the first complication of a spinal anesthetic—a spinal headache. Continuing his work with humans, Bier also came up with an approach now known as the "Bier block," now commonly used during surgery to repair carpal tunnel syndrome.

Dr. Harvey Cushing, neurosurgeon and the first physician to appoint another doctor to administer general anesthesia during surgeries he performed, is credited for naming the techniques posed by Bier "regional anesthesia." When local anesthetic is injected into the nerve that is closest to the operation, it is called a local block. This block is ideal for sewing up a laceration. Regional anesthesia, by contrast, is the practice of anesthetizing a specific region of the body by injection. For example, anesthetizing the entire foot for an operation on a toe, while the patient, who is usually sedated, remains conscious. Examples of regional anesthesia include epidurals for women in labor, spinals for numerous surgeries on the lower half of the body, and brachial plexus anesthesia to render arms insensitive to painful procedures such as carpal tunnel surgery, placing an arterio-venous fistula, and other upper extremity surgeries.

The Ineffectiveness of Animal Models in Anesthetic Development

Animal models are useless to the development of anesthetics for two reasons. First, reactions to gases and medicines vary, sometimes radically, between species. And second, the effectiveness of local and general anesthetic can be predicted based on their molecular structure. So, if all you want is an idea as to whether the chemical is going to be an anesthetic, look at its properties and structure. If you want more detail you have to study it in human tissue and humans. All species' brains do seem to respond similarly to the gases that anesthesiologists administer by inhalation to keep patients asleep (nitrous oxide, isoflurane, ethrane, halothane, and so forth). It takes about the same dosage of these gases to maintain slumber in a pig, dog, cat, or a rabbit as it does in a human. The difference between any two species does not usually vary by more than twenty percent. In other words, our brains are similar in their interpretation of painful stimuli and the ability of the gas to block that interpretation. This suggests that animals are good models, however in actuality they are not because of side effects. Though all mammalian central nervous systems respond the same to anesthetics, this cannot be said of the liver, heart and other organs. For example, methoxyflurane despite being tried on numerous animals, triggered kidney failure in humans. Fluroxene caused liver failure. Animal experiments predicted none of these potentially lethal side effects.[53]

Once researchers determine the effect of a substance, they still have the considerable challenge of ascertaining correct dosages. It makes no sense to use animals to determine dosages for anesthetic gases since anesthetic potency can be approximated by the oil/gas partition coefficient or lipid solubility. (Oil/gas partition coefficients measure how effectively a gaseous substance will penetrate an oil-based, or what is more commonly referred to as a lipid-permeable membrane, such as a human brain.) It requires only physical and chemical analysis of the gas to determine these dosages. The chemical structure of the gas itself indicates what will happen.

Narcotics, which sedate and relieve pain, also affect different species dissimilarly. If you have ever undergone surgery, you probably received a narcotic like morphine for pain control. If, during its development, morphine had first been tested on animals, real confusion might have resulted since it excites cats and causes convulsions in mice.[54] Happily, morphine was put to use as a narcotic before requirements for animal testing were instated. Like many narcotics, morphine slows breathing rates in humans and can, in overdose, result in death. But it affects other species' respiratory systems differently. In many animals morphine has no effect whatsoever on breathing. In fact, some animal-tranquilizing darts are made from very potent narcotics. The animal falls asleep, but its breathing is unaffected. M-99 is the narcotic that is used to sedate wild animals, but a single drop is lethal to humans.

The lack of parity between species in response to drugs used in anesthesia is so radical that it is impossible to form any conclusions whatsoever about human response based on animals. There exist variations between strains, breeds, and even genders of the same animal. For example, curare and pentobarbitone are more powerful in female rats than males. Alpha-2-agonists, such as xylazine, are ineffective in pigs but effective in small ruminants, such as goats. Xylazine works at much lower doses in dogs than cats, and induces vomiting in such a high incidence in cats that it is sometimes used as an emetic. This side effect does not occur in other animals. For dogs, xylazine is so successful that it is one of the most common canine anesthetics. It is not used in humans.

Ketamine makes an effective sedative or general anesthetic, depending on the dose, in humans, while in rats and guinea pigs it does next to nothing. In sheep and goats, it causes tremors and rigidity. Atropine is frequently used in humans as a drying agent, resulting in that parched mouth sensation after anesthesia. It makes sheep and goats salivate profusely. The general anesthetic propofol might have been banned if it had been tested on cats, because it can cause Heinz body anemia and death in felines. Yet human patients who receive propofol wake up after surgery quickly and with minimal nausea and vomiting.[55] In the *British Journal of Anesthesia,* a scientist added,

Other commonly used human anesthetic regimes are ineffective in some animal species. For example, midazolam and other benzodiazepines do not produce unconsciousness in dogs and cats and may cause agitation and excitement when administered by I.V. injection.[56]

To summarize, some animals can be anesthetized with the same medicines used on humans. However, the side effects can be markedly different from humans. Conversely, some animal anesthetics cause severe side effects in humans. The second Royal Commission on Vivisection stated in 1912, "The discovery of anesthetics owes nothing to experiments on animals."[57]

Rising Dissatisfaction with Animal-Modeled Surgery

During the early years of anesthesia, and following the implementation of sterile technique by Lister and others, the debate over the usefulness of the animal model escalated. Experimental physiologists, such as Claude Bernard, focused on laboratory validations, but already many surgeons advocated abandoning animal models. They deemed it not only futile, but also distracting from more relevant research. This was before litigious trends made animal models such a safe haven. Among antivivisectionists was one of the finest and most innovative surgeons of the time, Lawson Tait. He and others expressed concern that animal experimentation had done more harm than good because it diverted the focus from more reliable data.[58,59,60,61] Tait felt that animal experimentation should be stopped "so the energy and skill of scientific investigators could be directed into better and safer channels."[62]

Among these channels was the autopsy. One of the early revelations resulting from autopsy was that of appendicitis. Imagine a time when most illnesses were entirely unexplained. Seemingly healthy patients might suddenly become acutely and agonizingly ill. When they had horrendous abdominal pain on the right side, then dropped dead, no one knew why. A doctor named Fitz performed autopsies on 257 patients who had died of the same complaints and on 209 who had died of different complaints. By 1886 Fitz realized that the appendix could become inflamed and rupture. Animals would never have provided an explanation for those deaths, as ours is the only species that suffers from appendicitis, and most species lack an appendix.

Fitz's autopsies provided the basis for modern appendectomy. He recommended an operation to remove the organ. The idea was quickly adapted, first by Tait, and mortality from appendicitis plummeted. The appendectomy was a turning point. Since then, autopsies have proposed innumerable other surgeries including surgery for the correction of heart defects in babies, operations on the aorta, and operations to remove pancreatic cancer.[63] Lawson Tait went on to perform the first operation for a ruptured ectopic pregnancy in 1883, multiple oophorectomies and the first cholecystectomy (gall bladder removal) in Europe, the second in the world. His accomplishments greatly enriched this fecund time in the development of surgery. In the 1890s, the first splenectomies saved the lives of trauma patients who lay bleeding to death because of a ruptured spleen. With general anesthesia to make the operation possible, the first intestinal intussusception was surgically corrected in a two-year-old girl.[64,65]

Even while great strides were occurring in surgery, based on human data, controversy over animal models escalated. Animal experimentation was a growing industry, yet those who did not rely on its profits expressed serious reservations. Respected surgeon of the time, Sir Frederick Treves, stated, "My [animal] experiments had done little but unfit me to operate with the human intestine."[66] The pitfalls were echoed by the president of the Royal College of Surgeons, Dr. Moynihan, who remarked:

Has not the contribution of the [animal] laboratory to surgery of the stomach, for example, been almost negligible when it has not been potentially dangerous because divergent from human experience and therefore inapplicable.[67]

Innovations in Technology—Tools of the Trade

The industrial revolution ushered in technology to refine surgical skills, and enhance disease diagnoses. As more and more technology became available, more dedicated training was needed in order to use it effectively. This resulted in specialization. General surgery divided into neurosurgery, urological surgery, ophthalmology, otorhinolaryngology, plastic surgery, orthopedics, vascular, thoracic, cardiac, transplant surgery and other sub-specialties. This division, in turn, led to even more knowledge and further technological refinements, expanded exponentially by the electronics age. Surgery used to be about demanding stitchery. Now, breakthroughs in laser technology, computers, and fine-gauge equipment are transforming surgery into an art that would certainly have seemed miraculous not too long ago. It must be emphasized that these achievements have been the result of physics, chemistry, mathematics, computer science, and engineering.

Of the scores of revolutionary medical advances that are the direct result of technology, these are but a few:

- The concept of using electricity to coagulate bleeding vessels, electrocautery, was the aforementioned Harvey Cushing's invention.
- The invention of the ophthalmoscope, by Helmholtz, in 1851, allowed physicians to see inside the eye for the first time.
- The otoscope, designed in 1860, amplifies examination inside the ear.
- The laryngoscope, developed in 1854, allows better visualization of the larynx.
- Utilizing stains and dyes, plus the microscope, surgeons decide whether to remove tissue or treat more conservatively.
- Nuclear medicine is routinely used for evaluating cancers and hormonal levels of the lungs, endocrine organs and kidneys prior to and after surgery. Nuclear therapy is employed to treat hyperthyroidism and other diseases.
- Surgeons routinely place a catheter directly into blood vessels in the heart and liver, as well as other organs.
- The tonometer, the device pressed against your eye to measure intraocular pressure, allows early diagnosis of glaucoma.
- New suturing techniques minimize incisions in the eye, heart and other tissue. The development of the stapler has done away with sutures in some operations. This improves closure of the incision and decreases the operating time.
- Microscopic surgeries allow surgeons to reattach severed digits and limbs and save many people from lifestyle-changing injuries. The microscope is also used for the common procedure of placing tubes in a child's ears in order to decrease ear infections and hearing loss. Microscopic discectomy, the removal of a herniated disc in the back, is accomplished with the microscope. The microscope is also used in neurosurgery to allow removal of tumors without damaging delicate nerves. Microvascular surgery allows surgeons to work in very small areas, with minimal destruction of viable tissue. It is invaluable in some cancer surgeries allowing patients with advanced head and neck cancer to use some part of their larynx, saving their ability to speak.[68,69,70,71]
- Lasers have become a versatile surgical tool. The tiny size of this technology minimizes the destruction of viable tissue. They are employed either as a minute scalpel, or recently as an intravascular scalpel. Endovascular photo acoustic recanalization (EPAR) laser has been tested on twenty-six patients and found safe and effective. It is also faster then traditional thrombolytic therapy.
- Using a laser, ophthalmologists can correct potentially blinding conditions without anesthesia and on an outpatient basis. La-

sers remove birthmarks from the face and aid in stanching blood flow during surgery.

- Many procedures that would have required major surgery can now be performed nonsurgically via the endoscope. This instrument is commonly used to scrutinize the upper GI tract for ulcers or cancer and the inner colon for early cancers. A laser can be placed on the end of the endoscope and passed into the GI tract via the mouth or rectum. Cancer cells reflect the laser light differently then normal cells.[72]

- Surgical procedures are now performed less invasively with laparoscopes. Appendectomy, cholecystectomy, hysterectomy, hernia repair, kidney removal, and other surgeries were once only possible with a large incision and a prolonged recovery. Laparoscopes allow surgeons to make several very small incisions and insert instruments through them, thus decreasing the wound size, the likelihood of infection and other complications and decreasing recovery time.

- Tissue implants including artificial eyes, heart valves, penile prosthesis, skin expanders used in order to harvest more skin for skin grafts, artificial blood vessels, pacemakers and other advances came to us via technology.

- The gamma knife is used for brain surgery and vascular abnormalities and to section nerves involved in the transmission of pain.

- Acoustic microscopy, an extension of ultrasound, demonstrates internal conditions using sound waves without requiring dyes.

- Vacuum-assisted closure devices are used to prevent the skin necrosis from snakebite.[73]

- Electronic hearing implants stimulate the auditory nerve, which sends hearing impulses to the brain.

- Computerized canes, operating in conjunction with sonar, help the blind avoid obstacles and walk without being guided.[74]

- Impervious wound-edge protectors safeguard against postoperative wound infection.[75]

- Virtual reality flexible sigmoidoscopes help residents and other physicians train for live patient examinations.[76]

- Cryoablation is a new method of using an old concept. Extreme cold is used to kill pain-causing nerves and to kill cancerous tissue.

- Time-reversed acoustics employs a reversal of sound waves to destroy gall bladder and kidney stones.[77]

- Liquid oxygen, nitrogen, ether, ethyl chloride, Freon-12 and carbon dioxide are used to treat oral and skin lesions, brain tumors, Parkinson's disease, and neuromuscular disorders. These gases resulted from chemistry and physics and technology allows their use.

- Biodegradable microspheres injected into the brain safely and effectively deliver a sustained, local supply of a substance that helps radiotherapy after brain surgery.[78]

- Technology has revolutionized the most commonly performed general surgical procedure—the repair of inguinal hernias. Today prosthetic materials are used to repair the hernia. These materials are the result of advances in technology. In the old days, the hernia was simply sewn back together. This worked only short term and the hernia often recurred. Polypropylene mesh, polyester mesh, expanded polytetrafluoroethylene mesh, and polyester and absorbable hydrophilic collagen film are currently used today, providing better results than traditional repairs and decreasing the rate of recurrence. Mesh repairs also result in less disability and a quicker return to normal activity due to less tension on the muscles.

- Technology has also revolutionized the practice of anesthesiology. Today, patients are monitored with end-tidal CO_2 devices and machines monitor how much oxygen is in their blood on second to second basis.

Blood Transfusions and Blood Type

Throughout the history of surgery, and into the twentieth century, the subject of blood mystified the surgical community. Using animals in attempts to comprehend this marvelous life-supporting fluid did not light the way. Rather, it darkened and added unnecessary detours to the labyrinthine task of understanding. Given what we know today about blood's idiosyncrasies, it is hard to imagine the logic behind early blood transfusions wherein doctors actually infused humans with sheep blood. Since all the patients died, inevitably, the practice was abandoned.[79] Naturally, if human-to-human transfusions had been tried first, the chance of success would have been much greater.[80] But as we now know, even human-to-human transfusions demand blood type and factor specificity, as well as protection against infectious agents.

When Karl Landsteiner suggested ABO blood types as an explanation for reactions to blood transfusions in 1900, medicine got an indication that blood was far more complex than is visible to the naked eye. Landsteiner's remarkable discovery was based on clinical observation and test-tube research. Nearly forty years later, another layer of intricacy emerged. Many people still believe the human Rh factor was first described in rhesus monkeys. The true story is somewhat different. In the 1930s, Philip Levine's clinical experience with post-partum transfusion reactions in parturients drove him to *in vitro* examination of human blood. He found that some people lack specific antigens on the surface of their red blood cells. Those who are "negative" for these antigens produce antibodies against the blood of those who are "positive." Therefore, if what would come to be called "Rh-positive" blood is given to Rh-negative people, then a transfusion reaction can occur. For instance, when an Rh-negative woman gives birth to an Rh-positive child, she is exposed to the Rh-positive factor when some of the infant's blood crosses the placenta. This is irrelevant unless she needs a transfusion after birth and gets a second exposure. Then she may die. Levine's analysis suggested, in 1939, a factor additional to ABO typing, though it was not yet called the "Rh factor."

Landsteiner, who had won the Nobel Prize in 1930 for his blood typing discovery, repeated Levine's tests with rabbits and rhesus monkeys. Because Levine had neglected to name his earlier discovery, Landsteiner named the factor "Rh" after his primate subjects. All Landsteiner's rhesus monkey experiment did was to duplicate Levine's knowledge acquired earlier from human experience.[81,82,83,84,85,86] What the positive-negative factor is called is irrelevant. Further, more recent findings revealed that the Rh blood factor of Landsteiner, named after the rhesus monkey, does not even exist in humans. The antigen Landsteiner found in rhesus monkeys was not the same one described by Levine. Hence, the human factor mistakenly called "Rh" is entirely different from the one of the same name in monkeys.[87] (This was not Landsteiner's only monkey-based mistake, since it was he who suggested monkeys as models for polio, an error that delayed the vaccine by decades and resulted in a flawed vaccine that caused disability and death.) Historians have noted:

> Many years later it came to be realized that the rabbit anti-rhesus and the human anti-Rh antibodies are not the same. The vast literature that had accumulated made it impossible to change the name of the human antibody from anti-Rh, and the suggestion of Levine that the rabbit anti-rhesus antibody should be called anti-LW, in honor of Landsteiner and Weiner, has been widely adopted.[88]

Philip Levine, who imparted valuable human data about transfusion reactions that occur despite proper ABO typing, stated that Landsteiner's animal studies "contain nothing of clinical significance."[89] Despite these inaccuracies, even the federal government continues to insist that the Rh discovery came from animal experimentation.[90]

Our knowledge base about blood antigens and antibodies, as well as transfusion reactions, came from *in vitro* research and clinical observation. The same methods continue to add to our intelligence about hematology and diseases of the blood. We now know that not just a few variables, but scores of antigens can result in transfusion reactions. There are even antigens on white blood cells. Antigen discrepancies also lead to disease. Hereditary bleeding disorders in which the blood fails to clot adequately such as Von Willebrand's disease and hemophilia A were discovered by studying human populations and analyzing their blood in the laboratory.[91]

Regardless of what we now know about human blood's highly distinct character, the animal experimentation industry still attempts to find solutions via other creatures' plasma, platelets and blood cells. The historical failures of animal models with regards to hematology do not yet prevent the American Red Cross (ARC) from using animal models. The ARC owns the largest blood research test facility in the world—the Holland Lab located at Rockville, Maryland. Funded in part by charitable contributions, the Holland Lab uses mice, rabbits, and even toads in attempts to inform knowledge of human blood transfusions.

Fortunately, sane science, using more modern *in vitro* modalities, is greatly augmenting our understanding of blood, and blood disease. Research may even mitigate the need for blood transfusions. Transkaryotic Therapies of Cambridge, Massachusetts has developed a human cell line capable of producing the hormone erythropoietin. Erythropoietin increases the amount of red blood cells the body produces. In some cases, this eliminates the need for transfusions. Although erythropoietin can be produced from animal cells, the animal version differs from the human enough to create problems for the patient.[92] The manufactured form does not. Given this kind of quantum progress in therapies, would not the ARC and other research institutions do better to abandon archaic animal models and use modern-day research modalities like *in vitro* research instead? Thanks to *in vitro* science, researchers isolated the anticoagulant heparin from tissues and purified it. Afterward it was tested on blood and found to inhibit clotting. Though blood from animals was used, human blood would have provided the same and even more accurate results.

As researchers vie for tight monies to support their animal lab work and boast about its specious results, we forget how often mere observation provides sound insights. The usefulness of protamine, now used to reverse the effects of heparin, was witnessed in diabetics, who used it as a diabetic medication.[93] Coumadin is a longer lasting anticoagulant. Just as Paracelsus' chickens had serendipitously suggested ether's anesthetic properties, the idea for coumadin originated through animal observation, not experimentation. Cattle that eat sweet clover hemorrhage spontaneously. Analyzing the clover, researchers isolated the chemical dicumarol *in vitro*. They developed coumadin from this chemical.[94]

Animals do sometimes demonstrate an effect that later proves to have therapeutic value to humans. These random occasions tend to bolster the animal experimentation industry's argument. Nevertheless, as noted in chapters 1 and 2, such incidental associations that occur naturally are not scientifically sufficient to support reliance on supposedly predictive animal models. Animal-modeled results are not the same as natural occurrences. They are contrived and at best, hit and miss. At the worst, they are often dangerous. That does not, however, detract from the validity of testing an effect witnessed in animals in their natural setting on human tissue in an *in vitro* environment. That makes sense. It is scientific. We repeat, the effects of coumadin were an observation of a natural occurrence, similar to the observation that when humans took compound x and y happened. It was serendipitous. Like a drug being derived from observing reactions in humans. It was not experimental and does not lend weight to the animal model.

Radiology: to Operate or Not, and Where

Early surgeons were obliged to "dive in," unaided by today's growing repertory of fabulous diagnostic mechanisms. Not so now. Physicians have a visual capacity previously known only to Superman. With prescience and determination, outstanding scientists, mathematicians, and engineers have revolutionized medical observation with state-of-the-art medical tools. In radiological instrumentation, this amalgam of disciplines finds its zenith.

Wilhelm Roentgen's discovery of X rays in 1895 led, through refinement, to the thermionic vacuum tubes invented by William Coolidge a few years later. The outcome was diagnostic X-ray machines. These produced "radiographs" that penetrated to the body's interior through electromagnetic radiation. Solid matter such as bullets and bones showed up best. Over time, the diagnostic value of X rays was amplified, specifically through the use of dyes.

In the twentieth century, until the 1970s, progress in non-invasive medical scanning was slow. Since then, however, radiological and electronics developments have transformed surgery. Present-day virtual reality technology is giving surgeons not just X-ray vision, but micro X-ray vision, helping them minimize surgical wounds while avoiding damaging critical tissues. These highly informative procedures reduce the duration of the surgery and therefore the effects of anesthesia. They also permit operations that used to be far too risky. What led to this "image-guided surgery" and other visualization marvels?

In 1972, physicians' ability to appraise living systems suddenly improved by a quantum leap with the invention of the computer-aided tomography (CAT or CT) scan machine. Robert S. Ledley, a mathematical physicist, invented the first CT machine capable of creating cross-section images (tomes) of all parts of the body. The CT uses a series of X rays to build up a 3-D image of the body. Ledley later rendered the

CT significantly more accurate by reformulating the mathematical models on which it is based. These algorithms eventually contributed to development of MRI and PET scanners. Ledley received the National Medal of Technology in 1997.[95]

With the advent of the CT scanner, physicians could actually see tissues previously only visible during surgery. CT scanners offer clear views of any part of the anatomy, including soft organ tissues, with almost one hundred percent more clarity than a conventional X ray. Rotating 180 degrees around a patient's body, the device sends out a thin X-ray beam at 160 different points. Crystals positioned at the opposite points of the beam pick up and record the absorption rates in the tissue and bone's varying thickness. The scanner relays these data to a computer, which in turn transforms the information into a computerized image. These machines use no dyes and approximately the same amount of radiation as a conventional X-ray machine.[96]

An example of the tremendous diagnostic breakthrough offered by specialized CTs is the Focused Appendix CT scan (FACT). Appendicitis was always notoriously difficult to diagnose, with approximately one patient out of five undergoing the surgery without needing it. Another one in five was sent home from the hospital or physician's office only to have the appendix rupture, leading to life-threatening complications. FACT dramatically improved the odds of receiving a correct diagnosis. When radiologists place dye in the colon, it lights up the appendix, allowing more accurate diagnosis.[97] As only the right lower quadrant of the abdomen is scanned, this test is less expensive than a typical CT scan and certainly less than an unnecessary surgery or missed diagnosis. Another CT-related scan, the electron-beam computed tomography (EBCT) was designed specifically to detect coronary artery calcification in emergency room patients with chest pain.

Positron Emission Tomography (PET), developed in the 1970s, supplied a view of the physiology of disease. Brain physiology can now also be studied by a combination PET scanning and transcranial magnetic stimulation (TMS). TMS stimulates specific areas of the brain, while PET scanning monitors the portions involved in interpreting stimuli.[98] Mapping neural connections in a human brain utilizing both PET and TMS is revolutionizing our knowledge of how the brain works, with applications in neurology and psychiatry.

Many years ago, in 1938, Isidor I. Rabi discovered that atomic nuclei absorb and emit radio waves when placed in large magnetic fields. His discovery found immediate usage in laboratories, but not for medical visualization. Then, during the 1980s, Raymond V. Damadian built a prototype machine based on Rabi's work that could better visualize non-bone tissue. Working in a basement at Downstate Medical Center in Brooklyn, he created a large encircling magnet. This generated a strong, uniform magnetic field that interacted with radio waves to excite the nuclei. Its detection system picked up signals from the body and transformed them into a visual image. Damadian's technology was initially called nuclear magnetic resonance, but the word "nuclear" had negative connotations, so instead it became known as magnetic resonance imaging (MRI).

Finally, tumors that would have been otherwise missed showed up on MRI scans. Noting the different absorbent properties of cancerous tissue and normal tissue, Damadian published his results in 1971. Damadian did use mouse tissue to test his theory, but human tissue would have worked just as well. (In other words, the animals were not necessary.) Damadian's historic contribution to humankind was developed without grant money. After his success with the mouse tissue, he applied for NIH funding to build a human-size scanner. The NIH turned him down. In less than twenty years, this once unfundable project has become a mainstay of modern medicine. In 1988, Damadian received the National Medal of Technology.[99]

Since 1988, naturally, many enhancements have increased the efficacy of the MRI. MR angiograms determine blood vessel position. And "functional" magnetic resonance imaging (fMRI) shows which portions of the brain do what. Three-D models and real time scans help physicians to find the safest routes to their surgical targets, and in the case of tumors, to remove growths more completely. Trackable probes establish the trajectory and exact position of scalpels in correlation to the diseased area.[100]

However, some parts of the body are still difficult to visualize clearly. In development is a new medical imaging device that uses hyperpolarized gases to characterize the lungs. These gases provide an MRI signal that is as much as 100,000 times stronger per nucleus than those in conventional MRI. Though engineers were making good progress using human volunteers, the FDA decided to regulate imaging devices as drugs rather than as devices. Hence, developers were obliged to use dogs. About the only thing to come of using the dogs was that they barked at a higher pitch as a result of inhaling hyperpolarized helium.[101]

Magnetoencephalography (MEG) detects magnetic fields associated with brain activity such as occur in response to sounds and lights. It uses no X rays or radioactivity. MEG technology is far superior to experiments on animals with electrodes implanted in their brains. Electrical impedance tomography (EIT) is one of the very latest scanning technologies. Being small, mobile and cheap, it has a distinct advantage over large and costly scanners. Impedance imaging registers the body's electrical resistance that is affected in disease states.

Many are familiar with ultrasound in the context of the now routine sonograms of pregnant women. The U.S. Navy first employed ultrasound in 1955. John Julian Wild obtained permission from the Navy to use their equipment to measure the thickness of tissue and to aid in diagnosing the causes of abdominal pain. Based on his observations, he and an engineering friend built an ultrasound machine in his basement. A giant achievement, their invention and subsequent modifications proved invaluable for diagnosing tumors.[102] The technique provides diagnoses and therapies that would have previously required surgery. For example, sound waves sent back from a probe inside an artery convert into a cross-sectional picture of the artery wall, used to detect thickening or blockage. Ultrasound can also monitor the internal effects of new treatments, such as changes in blood circulation. It also aids surgeons during surgical procedures.

Another form of upgraded ultrasound is endoscopic ultrasound, an endoscope with a high frequency ultrasound attached. This allows the physician to penetrate body cavities and ultrasound the surrounding structures non-surgically, achieving not only visualization of the anatomy but analysis as well. It is used for evaluating the upper GI tract, the lower colon and rectum, the stomach and pancreas.[103]

Pro-vivisectionists often claim that engineers relied on animals to test their high-tech medical inventions. Yes, they sometimes did test on animals, but only gratuitously, not out of necessity. Even as new materials combined with high-tech instrumentation to reshape surgery over the last 150 years, animal labs, with all the commerce that accompanies them, thrived too. Ready supplies of animals, peddled by eager vendors, proliferated. Hence, innovators of technology may have employed animals unnecessarily. On no occasion did their use inform the development process in ways that could not have been garnered with human tissue or other human-based research. This point is critical.

Throughout the book we refer to research on human tissue as a more accurate than animals or animal tissue. Importantly, humans do not sacrifice this tissue for experimentation. Some tissues such as blood, placenta and umbilical cords are readily available. They are harvested from patients in biopsies or as waste (cosmetic surgery, amputations, mastectomies, and circumcisions). Cadavers are another source of tissue. Scientists continually improve techniques for keeping different tissues alive and functioning outside the body. Due to the plentitude of lab animals, human tissue is still an underused research material. A national protocol for achieving more efficient use of available human tissue would make these tissues more abundant and more available. Because the opportunity to use human tissue exists—whether researchers avail themselves of it or not—the argument that engineers tested on animals is not relevant. Each time we credit technology with progress, we are crediting human ingenuity, in chemistry, in physics, in math, in mechanics, and in electronics.

These triumphs—sterile technique, anesthesia, blood transfusions, radiology, human-based surgical observation and technology—transformed surgery into a practice of ongoing refinement. The possibilities today seem limitless, as though a time when literally everything will be able to be repaired is not far off.

Ophthamology and Orthopedics

Surgery increased in scope and elegance into an immensely respected field. Developments in anesthesiology, sterile technique, radiology and technology greatly refined procedures, boosting the level of repair and decreasing the damage. Many lives have been saved as a result of non-surgical advances in the field of surgery. Progress in research and surgeries gradually revealed how complex are individual systems within the body, and complex techniques grew to restore them. The inevitable response to this complexity was subspecialization. Space does not permit an examination of all the surgical subspecialties but it may be useful to consider two: orthopedics and ophthalmology.

Orthopedic surgery

Orthopedic surgeons are like carpenters. Their instruments are very similar to those used in construction—saws, hammers, nails, screws and other such items. It is difficult to find a field of medicine where technology has left more of a mark. For example, plaster casts were employed in ancient times to reduce dislocations.[104] New materials have rendered casts much lighter and even waterproof. High-tech engineering resulted in the mechanics of bone plating for fracture repair. Today, some patients can even use ultrasound bone growth stimulators at home to help heal long-standing fractures.[105,106]

Joint replacements are another mechanical wonder. As we age, chronic stress on our joints decreases mobility. Arthritis can set in and make normal motion impossible and painful. Efforts to replace arthritic hip joints took over a century to perfect.

Pioneers in hip arthroplasty used animal data to calculate the friction coefficient in the human joint space. As a result, they fashioned a lot of clumsy, noisy ill-fitted joint replacements. Sir John Charnley was an orthopedic surgeon during World War II. By chance after the war, a hip-replacement recipient happened into Charnley's clinic, sounding like the Tin Man without his oil can. Charnley's examination of her changed the way he looked at total hip replacements. The friction between the head of the femur and the hip socket was exaggerated. Consequently, her hip squeaked. Based on this and animal experiments that he did to "validate" his findings in this patient, Charnley decided to invent a prosthetic hip which self-lubricated. The body's natural joint fluid could not lubricate the new man-made hip. Surgeons had already tried numerous insular materials, including a pig bladder, in an effort to find a material that would support the body and not degrade.[107]

Needing materials that slid against each other with minimal friction, Charnley first tried Teflon. It worked well on animals and initially on humans, but failed after about one year. He placed three hundred new Teflon hips but none worked as animal models had suggested they would. The esteemed physician needed a different material. A salesman brought a new substance by Charnley's lab and suggested he try it. Charnley was not impressed, but his lab assistant decided to try it anyway. He found that the new material withstood use much better than Teflon. Furthermore, naturally occurring joint fluid could lubricate this new material. Charnley still required a substance to affix the prosthesis to the bone. Acrylic cement had been used in dentistry for decades, he learned, and it proved adaptable to joint replacements. Finally, all the necessary components were in place.

Charnley, who also contributed to the prevention of post-operative infections and postoperative blood clots, was quite a contrast to today's fame-seeking, grant-hungry researchers. Charnley turned down numerous positions of prominence in the medical hierarchy, seeking instead to continue his research and perfect his surgical skills. He derided younger surgeons who occupied their leisure with golf, suggesting that they should study woodworking with a lathe, as he did, in order to improve their dexterity and surgical skills.

Whereas vivisectionist Claude Bernard had claimed that the laboratory is the "true sanctuary of medical science," Charnley believed that "the true laboratory is the operating theater and the patient."[108,109] What he meant is that each time a patient undergoes an operation or takes a drug he or she is, in a sense being experimented on. No matter what one learns in a lab, it is meaningless because the human patient is always greatly different from the lab animal and may even be slightly different from other humans. Heirs to Charnley's innovation who experimented in animals were discouraged by the lack of parity. This shows up in their literature.[110]

Importantly, not all orthopedic implants work for all people. For years, Asians undergoing orthopedic surgery have experienced relatively high complication rates because the implants were designed for Western skeletons, which are bigger and longer and set at slightly different angles. Given these differences between members of the same species, is it not apparent that using other species as orthopedic models is wasteful?[111]

The use of cement for artificial joints led to other innovation. Bones in the vertebral column frequently break due to compression fractures from osteoporosis, metastatic disease, trauma, or hemangiomas. These vertebral fractures are very painful and difficult to treat. You cannot put a cast on a backbone. Surgeons decided to put cement in the fractured bone, inventing the vertebroplasty. Polymethyl-methacrylate is injected into the area of the break thus stabilizing the bone and relieving the pain.[112]

Arthroscopic surgery was another watershed event. Kenji Takagi first performed arthroscopic surgery in the 1930s, but his techniques did not catch on until the 1960s. With the development of fiberoptics that allowed smaller and brighter equipment, the field took off. Inserting small cameras through tiny incisions, surgeons can acquire data with very good image resolution via a small, lighted tube. The same technology is used in endoscopic surgery, thorascopic surgery and many other procedures. Over 1.5 million arthroscopic procedures are performed every year. The recovery time is much shorter than with the traditional surgery. Interestingly, it is still difficult to perform arthroscopic surgery on the knees of dogs because of the difference between dogs and humans with regard to the angle of the leg bones at the knee.

In vitro research has also helped advance treatments for orthopedic diseases. Cartilage provides a smooth surface in the joints for the bones to move across. Arthritis and other disease processes erode cartilage, leaving the bones to grate against each other. In 1994, physicians injected human cartilage grown in cell cultures into twenty-three patients suffering knee disorders. Since then this cartilage replacement procedure, which uses an arthroscope, has been performed over three hundred times. Surgeons inject cartilage into the knee through an arthroscope, with good results.

Meanwhile, attempts to extrapolate from animal experiments of the same nature have been frustrating. This is not surprising as the very formation of bone itself is variable between species. *The Lancet,* reporting on observations of bone transport, stated that:

> . . . the relative contributions of periosteal and endosteal callus [new bone growth at ends of fractured bone] vary between species: plugging the medullary canal has little effect in rabbits, but pronounced effect in dogs . . . since the results of these animal studies were inconsistent, it seems that that only way to discover more about the cellular events in the new bone was to study the process in man.[113]

Osteoarthritis, which is inflammation of the joints, is the most common type of arthritis. Because there is not yet a cure, millions suffer and take medication to alleviate its painful symptoms. Two doctors, Jonas H. Kellgren and Sir Thomas Lewis, bravely conducted experiments on themselves in order to understand the nature of arthritis pain more fully. By injecting chemicals into their muscles, joints, tendons, cartilage and bones, they gleaned appreciation of the way pain does not always register at its origin. Obviously, this insight would have been difficult, if not impossible, to acquire from animals. They had metal wires driven through their bones in order to find out exactly when a bone becomes sensitive to pain. Their self-experimentation produced very valuable data used today in the practice of medicine.[114]

In the early 1930s, British epidemiologists undertook the first study of arthritis in which participants underwent a clinical exam and answered a questionnaire. Thanks to this and subsequent studies, clinical observation and radiographic analysis, the risk factors for arthritis are pretty well defined—age, diet, weight, occupation, sex, crystalizing such as occurs with gout, and injury to the area. Heredity may also play a role.

Arthritis affects all vertebrates. Notably, however, humans alone have an upright posture and the associated movements are unique to humans. Considering the strong correlation between joint use and the disease, differences in animals degrade the relevance of data gathered in animal experiments. Or as Roger Lewin stated in *New Scientist*:

> Big species are not small species scaled up in a straightforward way— their bones become proportionately shorter and thicker as body mass increases, their metabolism slows down, as does heart rate, they live longer, mature later and have fewer offspring. . . . It is hardly surprising that being an elephant is not simply like being a mouse scaled up almost 200,000 times.[115]

Rheumatoid arthritis (RA) is a chronic autoimmune disease that results in inflammation of the joints and often deformity. Animal models of rheumatoid arthritis have been sought and employed in research for decades, but, as in other complex human conditions, no animal model is identical to the human disease.[116] As millions suffer from rheumatism, there are aggressive attempts to model the disease in laboratories. In spite of this, rheumatism has proved entirely inscrutable from the lab animal perspective, as a textbook on animal experimentation admits, "Unfortunately, there are no perfect nonhuman animal models for the study of rheumatic diseases."[117] Richard P. Carlson and Peer B. Jacobson of Abbott Laboratories further outline the problem of using rats to model humans:

> The majority of pathological changes seen in the AA and 10S PG-PS polyarthritis models [rat models of RA] closely resemble those described in the joints of patients with RA. However, the events in these chronic animal models occur over a short time period (0–50 days) compared with humans (0–50 years). Also, these models do not exhibit lymphoid follicles within the pannus, which contribute to the formation of intraarticular immune complexes. These immune complexes, which are derived from the interaction of rheumatoid factors (IgM/IgG) with aberrant antibodies from pannus follicles, exacerbate joint inflammation by activating complement, increasing neutrophil influx into the joint space, and inducing phag-

ocytosis by neutrophils which triggers the release of additional inflammatory mediators. RA also differs from AA and SCW polyarthritis in that flares and periods of quiescence occur in RA over the 30–50 years of the disease. This is not observed in the rat AA or 10S PG-PS SCW models, even in highly susceptible strains, where the disease progresses to end-state severe ankylosis/ossification 40–50 days post antigen. Finally, rat AA displays periosteal new bone formation, balanitis, some uveitis, and spondylitis characteristic of Reiter's disease (reactive arthritis) in man; these are absent in RA to the most part.[118]

What this is saying is, symptom-wise, the rats closely resemble rheumatoid arthritis patients. But they come down with the symptoms over fifty days, not fifty years. And within the joints the models do not exhibit the same growths that exacerbate joint inflammation, nor the release of inflammatory mediators as humans. Nor in the rats does the disease flare then ease. Finally, rats display new bone formation and other characteristics of an entirely different human disease called Reiter's.[119] These are huge differences. Only someone with a vested interest in animal models could say this "closely resembles" the disease in humans.

Interleukin II, a drug modeled after a protein in the human immune system and developed as a cancer therapy, has shown some benefit in regulating rheumatoid arthritis. *In vitro* research on tissue obtained from rheumatoid arthritis patients undergoing joint replacement surgery led to this finding.[120]

Osteoporosis is a loss of calcium that reduces bone density, resulting in brittle bones, prone to breaks. Hip and vertebral compression fractures are common. Hip fractures in our population have a twenty-percent mortality rate. That is, one in five of these patients die of related complications such as pulmonary embolism or pneumonia. The disease occurs mainly in post-menopausal women. Some believe spayed dogs should be good models for post-menopausal women since they are in such abundance, however they do not suffer from osteoporosis.[121] In fact, even experts in orthopedic research admit the "full parallelism of human symptoms with any single *in vivo* animal model [of osteoporosis] does not exist."[122] Until recently, women had to undergo X rays in order to monitor their bone loss, the frequent radiation bearing its own degrading impact. But the FDA has approved the use of an ultrasound device, which is not harmful, to measure bone loss in women suffering from osteoporosis.

Epidemiological, clinical, and laboratory studies alone account for the advances in treatment and diagnosis of osteoporosis. These studies have shown that vitamin D and calcium are important in prevention, as is a diet that restricts calcium waste. A high protein diet has been found to waste calcium, leading to osteoporosis. Smoking, alcoholism, lack of physical activity and certain medical conditions also predispose to the illness. Awareness of these causes can, of course, mitigate the effects of the condition as women alter their diets and lifestyles, exercise, and increase their net gain of calcium.

Gout is another disease that sends people to orthopedic surgeons. Gout occurs when the body metabolizes excessive amounts of uric acid and monosodium urate crystals accumulate in the joints. The inflammatory reaction these crystals cause leads to pain and swelling. Von Leeuwenhoek, famed for his work with the microscope, first observed the crystals under a microscope and described them. Naturally occurring gout does not exist in most mammals, but is rather common in birds and reptiles. In 1962, Rodnan injected crystals into the joints of humans and dogs to reproduce the symptoms. Why dogs were used is unclear, since they added nothing. One treatment for gout is colchicine, derived from the meadow saffron plant.

Specific orthopedic research venues that rely on nonanimal research methodologies for reliable data include:

- Cell cultures using osteoblasts (bone cells)
- Cell cultures using fibroblasts and myoblasts to study phenotypic expression and titanium biocompatibility
- Organ cultures using bones to study antibiotic toxicity, pH effects, and estrogen effects
- Cadavers for fracture fixation, spine stabilization, and ligament reconstruction
- Computers for prosthesis development, implant positioning, and Walffi's law
- Epidemiology for retrospective clinical studies and testing treatment hypotheses
- Physical models for studying the wear on joints and exercise physiology
- Bacteria to study calcification mechanisms and osteomyelitis pathogenesis[123]

Believe it or not, Barbie dolls are helping amputees. Jane L. Bahor of Duke University is taking the legs and knee joint out of old Barbies and building new fingers around it. The Barbie leg and knee joint is flexible and strong enough to support the activities of a normal finger and is allowing amputees to perform normal functions they hitherto could not perform. Foam and other material form a normal looking finger around the leg and knee joint. The old wire-constructed joints were not as aesthetically pleasing or functional as the Barbie-derived joints.[124]

A new pediatric leg implant allows better range of motion and extends as the child grows. This will obviate the need for additional surgeries to extend the device thus surgery will be necessary only to remove or replace the device.[125]

Serendipity continues to play a role in orthopedic discovery. An Orthovita scientist discovered a compound capable of increasing the speed at which broken bones mend. Orthovita's *Vitoss* is a porous material made from calcium that allows bones to grow through it. The body absorbs it, unlike steel rods or other hardware currently used. Therefore, the patient does not have to undergo another surgery to have it removed. *Vitoss* may also replace some bone grafts. *Vitoss* was discovered in a scene reminiscent of Alexander Fleming. The scientist was experimenting with the material on a hot summer day when he went inside for lunch. When he returned he noticed the material was dry and spongelike, full of small holes. The holes turned out to be the right size to allow bone to grow through them.

Ophthalmology

Ophthalmology is an ancient discipline. The first operation to remove cataracts was performed three thousand years ago. This is still one of the most commonly performed surgeries in our country, since approximately half the population greater than age sixty-five suffers from cataracts.[126] The operation was first described in detail based upon human observation in the nineteenth century.[127] By the early twentieth century, ophthalmologists could successfully perform the procedure with minimal risk to the patient. Note what the *American Journal of Clinical Nutrition* published concerning animal models of cataracts:

> . . . appropriate animal models for slowly developing age-related lesions [cataracts] are often unavailable. Furthermore, the environmental and lifestyle factors that may modify the influence of nutrient actions on the development of degenerative conditions cannot be sufficiently modeled in animal experiments.[128]

Cumulative exposure to sunlight has a demonstrated link with cataract risks. Epidemiology disclosed this and other factors that lead to cataracts, for instance a vitamin C deficiency. Animal experimenters co-opted the vitamin C data and attempted to reproduce it in animals. They deprived guinea pigs and rats of vitamin C and exposed them to UV radiation. Sure enough, the animals suffered more lens damage than the animals not deprived of vitamin C. But did this prove anything? They simply replicated already known human data and did not even do that adequately. Further, lens damage not cataracts occurred in the rodents. Many believe that lens damage leads to cataracts but this study did not produce cataract formation.[129] The *American Journal of Clinical Nutrition* goes on to state, "There are no animal models of senile cataracts in species that require vitamin C (as humans do)."[130] In fact, there are no predictable animal models of senile cataract formation. Too bad allocations for rodent studies did not go to public education about preventing cataracts by modifying diet and lifestyle and encouraging annual eye exams.

The ability to grow the human lens *in vitro* opens new horizons. Now researchers can watch the cellular processes that precede cataract formation, obviously a much more fundamental inquest. The study of protein metabolism, ion channels, and amino acid metabolism on the human lens are now possible as a result of this *in vitro* work.

Not surprisingly, the causes of many ophthalmological problems emerge when known events precede the disease. We do not need to damage animals' eyes to elucidate them. Amblyopia, for example, develops in children whose eyes are exposed to abnormal stimuli in infancy. Amblyopia develops in children when the eye is somehow exposed to abnormal stimuli. Such predisposing conditions would include, squinting, cataracts, poor uncorrected vision, and so forth.[131] One eye is weaker than the other. If the predisposing condition is not corrected, the child may grow up to be vision impaired permanently. The condition is usually caused by a defect in just one of the eyes and is confined to early childhood. So much was known in 1911, entirely secondary to clinical observation.[132] In these cases, the treatment has remained unchanged for one hundred years.[133]

Ophthalmologists usually use patches on their patients in an attempt to stimulate vision equally in both eyes. One treatment, to cover the good and then the bad eye alternately, was also developed in the early 1900s. It was first used in 1743. Other methods of treatment included covering the good eye for periods of time each day, thus making the weak eye work.[134] Often, since the children may have no experience of normal binocular vision, and therefore no developed neural bases, it takes quite some time for vision to correct itself.

Knowledge and treatment of amblyopia are a result of astute clinical observation, though vivisectors credit animal experiments. Animal experiments have repeated the results, but not led to treatments. Under oath, a pediatric ophthalmologist and full-time Senior Associate in Ophthalmology at Children's Hospital, Boston, and Assistant Professor of Ophthalmology at Havard Medical School stated:

> I do not believe that straining to find new ways of depriving cats of visual input has added or will add to our knowledge about the connections of the eye to the visual cortex in cats. In my opinion, the research reported here is essentially a trivial reworking, with minor variations, of work that has already been done. Second, even if it adds a little to our knowledge of visual connections in cats, the applicability of this knowledge to human amblyopia is essentially nil . . . .Clinical research, done with children who are actually suffer-

ing from amblyopia would seem to be the only way to find out more about how to treat this important condition which affects about two percent of the population.[135]

At Yerkes Regional Primate Research Center, researchers have experimented with eye patches on monkeys, but because of differences between species no useful data emerged.[136] One reason for these frustrated efforts is the vast differences between species in the parts of the brain governing sight.[137,138,139,140] The reasons why animals are not good models for amblyopia also lie in the species differences of the eye. Movshon and Van Sluyters state, " . . . there seems to be substantial differences among species in the degree and manner in which the environment affects visual development."[141] Despite all the money given to research to study eyes of cats and other animals, the *British Journal of Ophthalmology* stated in 1995 that, "Children with amblyopia are commonly given eye patches, which many hate wearing. Yet, 250 years after the introduction of eye patches ophthalmologists still do not know how long they should be worn or even how effective they are."[142] Only human-based data will change this.

Clinical experience with patients with known history, amplified by epidemiological surveys, has shown how poor vision correlates to poor diet (high cholesterol and inadequate fruits and vegetables) and smoking. These studies have informed our knowledge of diabetic retinopathy and macular deterioration (MD).[143] This valuable human data about human health does far more to mitigate the incidence of problem vision than experiments on animals.

MD, in which the central part of the retina (macula) deteriorates, results in blindness among the elderly, over eleven million people suffering in the United States alone. Age-related macular degeneration (AMD) is the leading cause of vision loss in people past their fifties.[144] Though there is no known cure, antioxidant pigments—specifically lycopene, lutein and zeaxanthin—found in fruits and vegetables have been shown to decrease the risk of MD.[145]

In vitro research has shown that certain genes predispose patients to MD.[146] Now, with the ability to determine whether they have this gene, individuals can correct their habits, eat diets low in fat with ample fresh produce, and avoid smoking. By studying this population, perhaps a cure or other preventive measures can be reached. Genetic researchers have identified the genes responsible for numerous other diseases of the eye too. This information may eventually lead to genetic therapies that will repair the disorders.[147,148]

On the technological side, already an experimental eye treatment is reducing vision loss among people with age-related MD. In photodynamic therapy, the ophthalmologist injects a light-sensitive medication called *verteporfin* into the eye, then shines a red laser light into the eye to activate the drug. This appears to destroy diseased tissue and blood vessels.

Animal models have long misled ophthalmological surgical innovation. Radial keratotomy is a surgery performed to improve vision without glasses. The first radial keratotomies were animal experiment-induced catastrophes. Surgeons thought they had perfected the procedure on rabbits, but the first humans operated upon were rendered blind. The reason lay in the difference between corneas. The rabbit cornea is able to regenerate on both sides, whereas the human cornea can only regenerate on one side. Surgery is now performed only on the surface that can regenerate.[149,150,151,152,153,154]

Microsurgical tools such as lasers are refining ophthalmology. The in situ keratomileusis, which corrects nearsightedness, uses a laser to shape the cornea. Photorefractive keratectomy (PRK) also improves vision by reshaping the front surface of the cornea using a laser. In another high-tech coup, a European research team fabricated a movable artificial eyelid that blinks in unison with the natural eyelid.[155]

Corneal transplant surgeries are performed frequently for the simple reason that tissue matching may not be required. This is because of a phenomenon called "immune privilege" in which large parts of the eye are exempt from the immune system. Therefore, corneal transplantations are unaffected by the rejection that poses a considerable challenge to other types of transplantation surgery. This privilege only extends to the white of the eye, the sclera. Beyond that, normal immune function returns and tissue typing must be considered. Therefore, except in patients with damage to larger areas of the eye, corneal transplants work very well. The more minimal operation was originally performed in the 1930s with ninety-five percent success. Better storage media, the microscope, precise medical instruments, and steroids led to the success of the modern-day procedure. Human studies and *in vitro* research have led to measures for extending immune privilege to a greater percentage of the eye's surface, thus allowing more patients successful corneal transplants.[156,157]

Biomedical engineers have successfully constructed artificial corneal tissue from immortalized human cornea cells. They anticipate that these "corneal equivalents" will eventually be available for transplantation, but in the meantime, they are useful for experimentation as a substitute for unreliable animal models.[158,159] Animal eyes have been very poor predictors of what will happen when the human eye is exposed to a substance.[160] (Another artificial material—a collagen-matrix barrier called *Corrositex*—has been government-approved to use in safety and corrosivity testing as a substitute for the skin of animals.) Grafts made of amniotic membrane are being applied in ocular surface reconstruction.[161] In fact, amniotic membrane—the innermost protective layer surrounding the fetus—has been used as a surgical material for several decades. Successful applications also include biological dressing for burns and skin wounds and surgical reconstruction of artificial vaginas, repairing omphaloceles, and preventing tissue adhesion in surgeries of the abdomen, head or pelvis.[162]

With no shortage of physicians peering into patients' eyes over the eons, a comprehensive picture of knowledge about human eye disease is growing. Here are a few of the many conditions elucidated by clinical observation of humans not animals:

- Keith-Wagener-Baker classification describes the changes in the eye that occur when a patient suffers from high blood pressure.
- The severe form of headache known today as cluster headache was first called Horton's headache after Dr. B. T. Horton, who described the constellation of symptoms, including those affecting the eye.
- Patients suffering from multiple sclerosis (MS) often exhibit a unique eye finding called Rucker's sign. This sign was used for years as an aid in diagnosing MS.
- Lambert-Eaton syndrome is a disease that may affect the muscles of the eye.
- Kearns-Sayre syndrome and Carney syndrome are diseases affecting the heart and eyes.
- Hollenhorst plaque is a cholesterol embolism usually from the carotid artery to the vessels of the eye.
- Stickler syndrome is an inherited disorder.
- Dr. Henderson instituted an operation for patients suffering from eyelid retraction secondary to Graves disease, now called the Henderson procedure.
- The ability to properly fit contact lenses was aided by the Dyer nomogram.[163]

Surgical Training

Surgery is invasive. It can be only as good as the surgeon's skill for marrying technique and observation to distinct internal tissues. Training surgeons acquire this skill gradually, first by watching their mentors operate on humans, then by operating under the direct supervision of those same mentors. Hence, they learn what goes where, how physiological processes occur, which procedures work best, and what sort of anomalies might occur . . . in humans.

Some medical schools still mandate that students in their first two years of medical school attend the traditional "dog lab." In this lab, students learn how drugs affect a dog's heart and lungs as well as how to perform certain minor procedures such as chest tube insertion on a dog. Physiology teachers, who get paid to supervise this lab, justify it by saying that they would not want the medical students to place their first chest tube unsupervised in an emergency room. Who would? But this concern belies the way medical schools really work. Students, like the surgery residents, observe numerous chest tube insertions before ever doing one. When the time comes, they will not be alone but will have probably two or three physicians watching them to make sure they are performing the technique properly. Only after multiple insertions, under direct supervision, will the by-then intern or resident be allowed to do one alone. Also, there are many differences between inserting a chest tube on an anesthetized dog and doing the same to an awake human with different anatomy from the dogs. Treat humans like the dog and the procedure will result in serious complications.

Each year, there are twenty-eight million human surgeries in the United States alone.[164] Roughly speaking, that is surgeries in numbers proportionate to the entire population of California. It is safe to say that medical students and residents do not suffer from a dearth of real-life examples. So, why should they practice on animals? Would they be better because of it? In England, routine surgical practice on animals was eliminated in 1876, with only rare exceptions. Over half of all medical schools in the United States, including top universities such as Harvard, Yale, Columbia, and Stanford, have removed the traditional dog lab from their curricula.[165] However, since teaching the dog lab is why some physiologists are employed, it still often exists in other medical schools. And to what benefit?

There is such an ardent resistance to ceasing animal labs that few controlled studies of its real merits have ever taken place. Funding for such comparisons is virtually nonexistent. But when studies have occurred on rare occasion, there has been no evidence of advantage. For example, researchers at Tulane University examined complication rates for the then relatively new procedure of laparoscopic cholecystectomy, removal of the gall bladder with an endoscope. Complication rates between those who practiced on animals and those who did not were no different.[166]

There are now many ways to practice an operation without using a living human or animal. Simulab Corporation of Seattle manufactures simulators for teaching laparoscopic surgical skills. They also make simulators for learning chest tube insertion, cricothyroidotomy, pericardiocentesis and peritoneal lavage, complete with bleeding tissues. Physicians training to treat patients suffering from major trauma, such as gunshot wounds, severe car accidents, and so forth, frequently use animals. In part due to the cost of the animal, the Advanced Trauma Life Support (ATLS) courses are expensive. One hospital in Georgia halved their costs by eliminating dogs from the course and using human cadavers instead. Participants evaluated the course as equal or better than courses using dogs.[167] More recent studies revealed that the physicians taking ATLS prefer human cadavers to animals. Maryland Shock-Trauma Institute, one of the largest trauma centers in the United States, has been teaching ATLS courses with human cadavers instead of live animals, for nearly ten years, with excellent results and reviews.

This is not surprising. If you treat and operate on humans it only makes sense to learn on cadavers. One of the primary objectives of this course is to teach doctors to open an airway in critical patients, the first step in any life-threatening emergency. To give you an idea of how inept animal models are for these purposes, you should know that on cats, routinely selected for practice, the endotracheal tube is placed with the animal in 180 degrees opposite the direction in humans—a rather important difference.[168] Salvatore Rocca Rossetti, surgeon and Professor of Urology at the University of Turin, Italy, states that:

> Nobody has become a surgeon because of having operated on animals. He has only learnt wrongly through animals. I have been able to see this over my many decades as a surgeon, also as a director of hospitals. I have carried out tens of thousands of operations on people without ever performing them first on an animal.

Dr. Werner Hartinger of Germany agreed:

> The claim, frequently heard, that animal experimentation is vital for the training of surgeons and that practice on living animals is necessary to gain manual and operating skills cannot be left unchallenged. A surgeon acquires his basic knowledge by observing and then assisting his teacher. In time, according to his experience, ability and manual dexterity, he participates in supervised operating duties, until the surgeon responsible for his training decides as to when he can start operating on his own. Specialized knowledge of microsurgery is gained in the same way, just as working at the surgical microscope does not call for operating on animals.[169]

These esteemed physicians, who practice outside the United States, can speak freely because they are not part of America's pro-vivisection medical establishment. They will never be forced to ask a surgeon who practices in an institution that profits from animal models to testify on their behalf in a malpractice suit. Consider the scope of innovation, only partially described in this chapter, which has occurred despite misappropriation of research money to animal experimentation. Consider what surgeons could do if animal-model-tagged funds were diverted to technology in order better to educate and train our future surgeons. Both considerations augur tremendous progress, progress only restrained by ignorance.

chapter 6

Pediatrics

Reproduction studies have been performed in rats, rabbits and mice at doses up to six times the human dose and have revealed no evidence of impaired fertility or harm to fetus due to the drug. There are, however, no adequate and well-controlled studies in pregnant women. Because animal reproduction studies are not always predictive of human response, the drug should not be used during pregnancy unless clearly needed.

—*Physicians' Desk Reference*

Perhaps nothing is as disturbing to an adult as a helpless youngster weakened by disease, and the animal-testing community realizes this. Sick children are its most frequently used *ad populum* argument. By linking animal models to saving children—no matter how tenuous the connection—pro-animal model lobbyists become nearly invincible. When we lecture about animal models, the vivisection lobby invariably aims to depict us as people who prefer rats to children. This facile characterization, which skirts our argument, is often effective from the viewpoint of the uninformed onlooker. The truth is that the use of sentimentality is really only an attempt to compensate for a lack of science.

When these same avid animal-model supporters are pressed to produce *facts* regarding animal models' actual role, they are at a loss for documentation. Therefore, they renew their emotional appeal by referring to sick children. They can bring on the tearful parents and the child's physician. If he has inferred or told them that knowledge gleaned from lab animals saved their baby . . . well, say no more. That is all the convincing they need. The distraught parents repeat what their doctors assure them is true. Facts, such as those in this chapter, get blurred in this charged setting. Congressmen and public officials know how public relations work, and they certainly cannot go on record as being against kids.

That these already beleaguered parents are, in truth, callously manipulated by the medical research establishment goes unsaid. No one asks why these doctors support animal experimentation. Do they work for a university in which the NIH heavily funds research on animals? Have they ever taken a critical look at animal experimentation to see if what they were taught is true? Or do they, knowing no better, just parrot the adage that all great medical advances have come from research on animals?

To be fair, most doctors and researchers have no vested interest in animal models. They simply have never judiciously evaluated the situation. Why should they? Physicians train to be just that—physicians. They do not train to be medical researchers or historians. Most hear that animal experiments made the treatments they administer possible and never give it another thought. Working twenty hours per day during internship and residency does not lend itself to critical evaluation. Many of the physicians, parents, and patients who give testimonials to animal experimentation are merely misinformed.

On the other hand, there are people who are paid to mislead the public: spokespersons for the tobacco industry, spokespersons for the animal experimentation industry, and those who support themselves on the grant money for animal experiments. Here is an example of their handiwork. The Foundation for Biomedical Research, a pro-vivisection lobbyist group, produces a video that recounts the stories of numerous children saved as a result of lab animals. Even its title, *HOPE,* is designed to elicit tears from the tender-hearted. In sweeping generalities but without facts, the video assures us that the only way to save babies is through the use of animal models. It tops off with the death of one child whose life could have been saved with "more animal experiments."

Such scheming PR, which relies solely on emotional appeals, should raise this suspicion: if the facts of the cases are so persuasive and overwhelming, why do they not reveal them? We want to hear, truthfully, what animal models have done for sick children. Tell us from where the real advances issued. Without facts, vivisection does not make sense. Pick at the edges of the pro-vivisection argument and you find researchers opting for what is expedient in the short term (grant money and papers) in lieu of what actually works in the long term—human solutions.

In this chapter, we ask you to winnow away the emotional fears and look at the facts in an unbiased way. The diseases and conditions described in this chapter prey on our young people. But you will read that animal experimenters, under guise of aiding them, prey on them too. You will see how animal models have failed, at taxpayers' and philanthropists' expense, to alleviate children's suffering.

An impartial assessment of the influences working for and against our children is critical. There are 7.2 infant deaths per 1,000 births, a figure that would not be appalling if it were elsewhere than the most developed and technologically advanced country in the world.[1] Of the twenty-six industrialized nations, the United States ranks twenty-second in infant mortality. Although the United States spends more money per person on health care than any other country, only four developed nations have worse statistics than the United States. Moreover, the rate of premature births continues to increase here, while it decreases in other industrially developed nations. A major part of the reason for this is, as we shall see, that scientific resources go to animal studies here instead of human health care and education.

Pediatric medical research is a priority and it does demand money. But it is time to demand results from that money, results that are applicable, not to mice, not to baboons, but to human babies and children.

From Conception to Birth

Embryology, the study of the fetus, is the first place to begin our analysis of what the animal model has done to children. Researchers determined how human embryos develop by examining aborted fetuses from healthy pregnancies.[2] Examinations of embryos aborted at one week, two weeks, one month, and so forth construct an informed picture of human development, which is of course quite different from the development of other mammals that gestate at a different pace and take on other characteristics. Indeed, stem cells, which we discussed in the Internal Medicine chapter, offer insights into epigenesis that cannot be studied directly in humans or in animals. Technology is allowing us to learn more about the brain and its maturation. Embryologists employ functional MRI (fMRI) to understand how the fetal brain responds to stimuli. If more money had been put into creating imaging technology instead of animal models of fetal brain development, we could have had this option sooner.[3]

One of the many issues that often gets lost in the haste to drum up more experiments in animal labs is the importance of maternal and prenatal care to the infant. This is the most fundamental aspect of birthing healthy babies, and no number of animal studies is going to improve the way individual mothers care for themselves and their gestating babies. This is entirely a case for preventive health.

Vested-interest groups frequently claim that infant and mother mortality rates have plummeted over the last century because of animal models. This is misinformation. Mortality rates have plummeted, but not because of animal models. In the early 1900s, six to nine mothers out of a thousand died of complications related to pregnancy and childbirth. And ten percent of infants died before reaching the age of one year. By 1999 the infant mortality rate had declined by ninety percent and the maternal mortality rate by ninety-nine percent. According to *Morbidity and Mortality Weekly Report* the reasons for these declines are "environmental interventions, improvements in nutrition, advances in clinical medicine, improvements in access to healthcare, improvements in surveillance and monitoring of disease, increases in educational levels, and improved standards of living."[4] Note, there is no mention of animal experimentation.

Birth Defects

Birth defects are any abnormality manifesting at birth, due to several influences—the birthing process, genetics, premature delivery, exposure to teratogens in the womb, and so forth. (Teratogens, as we have seen, are birth-defect causing substances.) Birth defects, whether anticipated or actual, cloud many pregnancies. Every year in the United States, 150,000 babies are born with birth defects, and birth defects are the leading killer of infants. Studies published in 1991 reported that twelve percent of all hospitalizations of children are due to birth defects and genetic diseases. These children are admitted younger and stay in the hospital longer. Their hospital bills are over 184 percent higher than those of other children, and they are 4.5 times as likely to die than other hospitalized children.[5] Birth defects and disabilities cost eight billion dollars a year in health care, special education, and lost work, much of it borne by the families of these children.

Premature infants account for over half of all birth defects and the majority of neonatal deaths. Over eleven percent of babies are born prematurely. Real progress in treating the special needs of premature infants is the result of the institutionalization of neonatal intensive care units and the specialized equipment associated with them. Antibiotic therapies, special breathing apparatus, proper nutrition and hydration all have issued from clinical research and technical advance, not animal experimentation. Nonetheless, though our ability to care for premature infants meliorates, the medical system, and indeed our whole society, faces huge challenges because the premature birth rate continues to rise.[6] We do know, through epidemiology, what causes premature birth and how to prevent pre-term labor.[7] We need to put this knowledge to work with preventive measures such as education. Re-creating the effect in other animals does nothing.

With the medical cost of caring for children with birth defects now over one billion dollars annually, sound research is a priority. Between 1979 and 1989, the Centers for Disease Control and Prevention tracked thirty-eight birth defects. During this time period, incidence of twenty-seven of the birth defects actually *increased* in the U.S. population. Nine

did not change and only two decreased. Birth defects continue to proliferate, despite billions of dollars in animal-model funding, from both
public and private sources. It stands to reason: the more families know
about the sources of birth defects and childhood diseases, the better
equipped they are to avoid them. The role of prevention in pregnancy
cannot be overemphasized. Sometimes protecting an unborn child is
difficult, particularly when unknown exposures and genetic factors are
at play. However, there are measures every mother can take that are not
difficult. Prenatal care is an exigency, because it monitors the mother's
complex bodily changes and the fetus's progress.

One important aspect that prenatal care stresses is good nutrition.
Yet, it is too often overlooked since we adults ourselves can frequently
get by with inadequate nutrients. However, fetuses cannot. Studies indicate malnutrition in a pregnant woman predisposes her child to infections later in life. Good nutrition during gestation may program the
fetus's immune system to fend off infection, whereas poor nutrition
weakens the fetal immune system. Epidemiological studies, for example,
in Gambia clarify this. Records of births since 1949 revealed that babies
born to mothers who ate well while pregnant lived longer than those
born to mothers during the wet or "hungry" season. Obviously, the ill
effects of poor nutrition during pregnancy are not going to be cured by
animal experiments.[8,9]

Nutrition during pregnancy is not an easy chemistry. This is why
prenatal care is so important. Expectant mothers may develop gesta-
tional diabetes or hypercholesterolemia or any number of other conditions that may affect the infant. Maternal hypercholesterolemia, for
example, is linked to atherosclerosis progression in offspring.[10] Good
nutrition includes vitamins and minerals, of particular importance to a
growing fetus. In the early twentieth century, Roger Williams submitted
a paper to the *Journal of Biological Chemistry* on the discovery of
vitamins. The editor suggested that he undertake experiments on animals
in order to further knowledge of that area. Writing many years later in
1990 Williams stated, "If I had followed his advice many of the vitamins
we have known for thirty years might still be undiscovered."[11] Today,
clinical observation and epidemiology continue to establish the exigency
of proper nutrition during pregnancy. Epidemiology, for example, demonstrated that obese women have a two-to-fourfold increase risk of
having a stillborn baby.[12]

Everything we know about birth defects has emerged this way, from
human-based research—epidemiology, clinical observation, and *in vitro*
research. Together, these modalities have elucidated the causes of some
birth defects by demonstrating bad nutrition, inadequate prenatal care,
and/or common exposures to specific toxins. Interested parties see considerable evidence that women who gave birth to infants with similar
defects were exposed to the same toxic chemicals and/or did not take

care of themselves in the same ways during pregnancy. Other birth-defect causes we know since follow-up *in vitro* research uncovered genetic defects. Comparable studies of healthy expectant mothers result in data on the care required for healthy babies.[13]

Here is a case in point. Recently, epidemiology revealed an increased risk of death and birth defects in fetuses whose mothers lived on or near farms. Pesticides including phosphates, pyrethroids, halogenated hydrocarbons, carbamates, and endocrine disruptors were sprayed not far away. According to Dr. Irva Hertz-Picciotto, professor of epidemiology at University of North Carolina, "This is the first study to our knowledge of pesticides and pregnancy in which exposures were in close proximity to the subjects and the verification of pesticide use was objective, not relying on people's memories of what they might have been exposed to." Dr. Bell, another researcher who co-authored the study, stated, "The take-home message is that we did find an increased risk for women living near agricultural fields where pesticides were applied during the early weeks of their pregnancies, but these results are not conclusive."[14] Pregnant women exposed to organic solvents at work also increase the risk of fetal malformations, a Canadian study concluded.[15] Air pollution has also been linked to postneonatal infant mortality.[16] These are just three epidemiological analyses out of many that have identified hazardous exposures.

Epidemiology confirmed that exposure to cigarette smoke increases the incidence of middle ear disease, asthma, bronchitis and wheezing illnesses in children, as well as umbilical cord blood mutations.[17,18,19] Babies born to women who smoke, or to women exposed to passive cigarette smoke, can even have carcinogens and gene mutations that may lead to leukemias and lymphomas in their body at birth.[20,21,22] Another study of pregnant women proved that smoking decreased intake of vitamin C.[23] There is also evidence that maternal smoking can lead to childhood conduct disorders.[24] Repeatedly, epidemiology, not animal models, found these links.

We also learned about the hazards of organic solvents, lead and mercury exposure through epidemiology.[25,26,27] For instance, men whose occupations expose them to lead are at increased risk of fathering preterm infants.[28] Epidemiological studies recently revealed that, even in countries where mercury levels comply with World Health Organization guidelines, mercury still produces birth defects including developmental delays and mental retardation.[29] One source of mercury, discovered thus, was coal-burning power plants. That radiation caused cancer to the fetus was undetected in animal studies but verified via epidemiological studies between 1958 and 1962.

Still, with all this knowledge, one and one-third million expectant mothers receive inadequate prenatal care each year and remain ignorant about nutrition and exposures. Prevention is chronically under-funded while scarce funding dollars go to animal experimentation. In a typical example, cocaine exposure *in-utero* results in mental developmental complications. This is not news. Copious epidemiological evidence confirms it.[30] Prenatal cocaine exposure costs the United States more than $300 million annually.[31] One would think the system would support preventive measures, not hollow gestures. Nevertheless, researchers wanted to study whether mothers' cocaine consumption might actually lead to learning disabled offspring—in rats. The March of Dimes gave them $2.5 million. The scientists found that maternal cocaine exposure influenced the learning of female rats but not male rats. Interestingly, another study found male rats affected and not females. In the words of Dr. Anderson:

> It needs to be recognized that, as well as a personal tragedy for the families concerned, congenital abnormalities now present a major public health problem. *Evidence of an effect in rats is not also evidence of an effect in humans* [emphasis added].[32]

Learning disabilities caused by maternal cocaine consumption are a totally preventable birth defect, and no one needed rat studies to prove it. A quick trip through the neonatal intensive care unit to view "crack babies" would have offered all the insight needed about the hazards of cocaine to the unborn child. Rat studies do nothing for these poor infants. To offer a false choice between the life of a lab animal and the life of a child does nothing either. Dr. K. S. Larsson and colleagues stated in *The Lancet* in 1982:

> What is the value of routine tests in animals for prediction of chemical teratogens? The correlation between known effects in laboratory animals and clinical adverse effects is very low.[33]

How many poor mothers could have been given pre-natal care with the $2.5 million? A British study confirmed the rising incidence of congenital birth defects in children of women who use ecstasy (methylenedioxymethamphetamine).[34] Now that we know this, is it time to confirm with animal studies or make inroads into prevention?

Every year 12,000 babies are born with birth defects because their mothers drank alcohol while they were pregnant, and the prevalence of alcohol consumption among pregnant women in the United States is increasing.[35] Fetal alcohol syndrome (FAS), a high profile childhood affliction, was described clinically and linked to alcohol consumption epidemiologically. FAS gives rise to facial dysmorphology, growth retardation, central nervous system defects, and other deformities. Pregnant women who consume five or more alcoholic drinks per occasion, one or more times per week, have a 3.4 to 4.6 times greater risk of bearing a child with cleft-lip birth defects.[36] Drinking during the early months of pregnancy, before a woman may know she is pregnant, can lead to FAS, particularly cleft-lip facial deformities because the lip and palate form during this period. Heavy or even moderate drinking later in pregnancy can also have tragic consequences. We do not need drunken pregnant rats to tell us that FAS-derived problems are avoidable. Educating the public about the ill effects of drinking during pregnancy and enforcing policies to protect the fetus could totally prevent FAS.[37]

The environmental threat to unborn babies is not the same in all species. Daniel E. Koshland, editor-in-chief of the journal *Science*, stated:

> There are great similarities and important differences in the DNA repair system (DNA repairs keep cells healthy) as one goes from species to species. These differences explain why a chemical found to be carcinogenic for one species can have a smaller or greater effect on another species. Aspirin, for example, causes birth defects in rabbits, but is harmless in humans.[38]

In fact, our choice should be good science or bad science, the gathering of applicable data versus useless data. To repeat, all advances in treating and preventing birth defects have come from clinical observation, epidemiologic research, *in vitro* follow-up, technology, and advances in clinical care, not animal studies.

Spina Bifida, Anencephaly, and Neural Tube Defects

Spina bifida and anencephaly affect one in four thousand pregnancies a year. Neural tube defects occur in approximately one birth per thousand. Animal experimenters, by inculcating animals with teratogens, have attempted to model and prevent these diseases and defects, even though epidemiological evidence has proven time and again that they stem from folic acid deficiency. The model makes no sense. Efforts to re-create the effects of folic acid deficiency and the ensuing neurological defects in animals have of course failed. For instance, neural tube defects in embryos of curly tailed mice are not prevented by folic acid. Other mouse models of neural-tube defects do not respond to folic acid either. Animal models are either unresponsive or poorly responsive to folic acid. No one yet knows exactly how folic acid prevents neural-tube defects.[39,40,41,42]

Meanwhile, according to some reports, two-thirds of new mothers are still unaware that folic acid is important during pregnancy. Expectant mothers need four hundred micrograms of folic acid per day to avoid risk of their baby having a neural-tube defect. A recent survey found that one-third of all women surveyed had never heard anything about folic acid. Regardless of the exact number, there is plainly too little awareness of folic acid's ability to prevent birth defects or knowledge of the recommendations concerning folic acid. This is another example of how dollars might be applied to educating humans rather than inculcating animals with toxic substances.[43]

Prematurity and Related Disorders

Low birth weight (LBW) is an immense problem with costs to the health-care system in excess of two billion dollars yearly according to a study in 1990. LBW has been linked, through epidemiology, to cerebral palsy, retinopathy of prematurity, respiratory distress, autism, epilepsy, and other conditions. Women who receive little or no prenatal care risk delivering babies underweight and pre-term because they never learn that poor habits and inadequate nutrients work against their babies' well being.

Nicotine exposure *in utero* can result in growth retardation and pre-term birth, as the innumerable low birth weight babies born to smokers confirm.[44,45,46] Yet, there is no end to the research the March of Dimes has funded in which scientists pump nicotine infusions into cages full of pregnant rats to find out if it stunts growth in offspring. This is an utter waste of money, obviously, along the lines of the cocaine-addicted rat experiments.[47,48,49] Approximately twenty-five percent of all women smoke throughout pregnancy. In lieu of supporting animal experiments, why not put this money into smoking prevention programs for adolescents and pregnant women?

Cerebral palsy, broadly, refers to several nonprogressive motor function disorders, which occur from brain damage, often before birth. Epidemology reveals that its incidence increases in babies born with very low birth weights. The discovery of a mitigating factor—not a cause, but a potential preventive measure that would keep babies from being born ill—arose from clinical studies. Noting that pregnant women who receive magnesium for fluid retention had a lower incidence of mental retardation in their infants, scientists found that giving magnesium to mothers could reduce mental retardation in low birth weight babies by forty-nine percent and reduce cerebral palsy by sixty-three percent. This is a classic example of clinical research yielding very beneficial results. *In vitro* analysis of blood samples from children with cerebral palsy further elucidated its makeup. The studies showed higher levels of inflammatory and coagulation agents, which may be due to either genetic or environmental influences. This work points to avenues for treatment with anti-coagulants and other drugs.[50] Detailed urine analysis of newborns can identify which infants will go on to develop cerebral palsy, signaling the need for earlier remedial therapy.[51]

Premature babies do not have fully developed lungs, and often lack a lipoprotein complex called "pulmonary surfactant" that is necessary to prevent lungs from collapsing. The surfactant is produced late in fetal life in preparation for birth. Lack of surfactant can provoke a disease called respiratory distress syndrome (RDS); the babies' lungs either partially or totally collapse. Without supplemental oxygen, they may die. Dr. Mary Ellen Avery correlated surfactant and RDS during the 1950s by examining cellular extracts from human babies' lungs.

RDS can be treated with natural (animal-derived) or synthetic surfactant. One report analyzed eleven clinical trials and found the natural surfactants curtailed neonatal mortality, retinopathy of prematurity, and bronchopulmonary dysplasia better than synthetic.[52] A more recent study revealed no difference between natural and synthetic versions.[53] When deciding which preparation to give, the authors noted that the "minor theoretical risk of transmission of infectious agents in natural preparations" should be considered. The risk is not insignificant. Concern over implanting animal-derived products into humans is escalating.

The Truth about Teratogenic Testing

Karnofsky's Law states that any substance can be teratogenic if given to the right species, at the right phase in development, in the right dose. That means all medications can cause birth defects in some species.[54] To give you an idea of the sensitivity level, even common table salt and water are teratogens in some species when they are administered in the right dose at the right time.[55,56] An immense amount of experimentation has proven Karnofsky right.

Why are unborn babies so vulnerable to toxicity? Until thalidomide, the fetus was considered by many to be immune to adverse effects of drugs taken by the expectant mother.[57] We now recognize that, by their very nature, drugs alter living systems. A good drug will treat the source of a disease without disturbing overall well-being. However, no living system is as precarious as that growing *in utero*. The cells are dividing, and the body is growing faster than it ever will again. With all this activity the cells are more prone to the adverse effects of toxins than they will be after the child is born. Drugs that are tolerable in adults—such as thalidomide—can result in a baby being born with malformations. The cells are at maximum vulnerability at the fetal point.

The accuracy of Karnofsky's law makes it incumbent on drug manufacturers to conduct teratogenicity testing, as we reviewed in the Development of Medications chapter. And though it is unnecessary, unscientific and inaccurate, testing animals for teratogens continues. Despite its evident drawbacks, the convention is not illogical when viewed from a perspective other than the pregnant patient and her family. People are making money and companies are keeping from being sued.

D. F. Hawkins, Professor of Obstetric Therapeutics and Gynaecology at
Hammersmith Hospital, London, stated in 1983:

> The great majority of perinatal toxicological studies seem to be
> intended to convey medico-legal protection to the pharmaceutical
> houses and political protection to the official regulatory bodies,
> rather than produce information that might be of value in human
> therapeutics.[58]

Ironically, pharmaceutical companies rely on animal testing both to
force approval of drugs that will not necessarily be safe for consumption
by pregnant women, and to reduce or eliminate restitution in case pregnant women who have taken the drugs give birth to birth-defected
babies. (In *Sacred Cows and Golden Geese* and elsewhere in this book,
we elaborated on the drug thalidomide's disastrous impact on newborns
and the politics that led to mandated animal testing in drug development
with the passage of the Kefauver-Harris Act. This too was ironic, since
animals did not indicate thalidomide was dangerous. From then on, drug
companies have made testing on pregnant female animals an established
protocol.)

Animal testing gives pharmaceutical companies evidence of due diligence in the event birth defects do occur in humans. But it is specious
evidence, because clearly the due diligence is not diligent. Animal experimentation does not protect human babies. Even as early as 1963, only
two years after the Kefauver-Harris Act passed, *The Lancet* published
the following statement, casting doubt on the new protocol:

> In fact, the pitfall is that, having found no teratogenic effect in a
> 'sufficient number of different species of laboratory animals', one
> can still not be sure of the effects on the human fetus, which is
> always the ultimate purpose of investigation.[59]

Humans, pregnant or nonpregnant, furnish the only accurate information about a drug's effectiveness and toxicity in humans, but the pro-
animal model community claims that we need animals to mime metabolic processes, to observe how the body, any body, absorbs substances.
The problem is, "any body" cannot be matched to humans. All animals
metabolize differently. One can watch the argument in favor of animal-
modeled metabolism crumble during teratogenic experiments on animals. Diverse reactions to a specific teratogen are due to different rates
of metabolism, as well as to qualitative differences in metabolic pathways. The longer the drug stays in the system, the more damage it does.
For example, chlorcyclizine is a safe, useful antihistamine drug for humans. In rats, it is teratogenic because the steady-state level (the amount

of the drug in the blood at a given time) of the drug is three times higher than in humans.

The way the drug enters and moves through the system also determines its impact. In humans, drugs are metabolized in the liver and in other organs. Some drugs in humans are totally metabolized in the liver while others are not. Same in animals. Also if a drug is one hundred percent metabolized in the liver in humans, it may be ninety percent metabolized in the kidney in a rat. The non-liver sites are called "extrahepatic." Extrahepatic activity is negligible or absent in fetal rats, guinea pigs, rabbits, and swine. This greatly alters susceptibility to teratogenesis in those species. For instance, the drug imipramine metabolizes into different metabolites, depending on the species. So animals—which experimenters claim we need to stage metabolism—do not metabolize the same way. So much for that reasoning. *Nelson's Textbook of Pediatrics* states, "Much of our knowledge of fetal physiology has been obtained from animals and often is not directly applicable to man."[60]

The means by which any given substance enters the body also influences its teratogenicity. Drugs given by mouth may not be teratogenic because they pass through the digestive system, but those administered directly into the blood stream by injection may be. The medium that the drug is dissolved in is sometimes teratogenic in various species too. Likewise, combinations of drugs may prove teratogenic, whereas they are safe when given alone.[61] Further, the biochemical reactions by which substances cross the placental membrane are at variance between species. Placentas differ greatly between species. A substance that crosses the placental barrier to affect the fetus in one species may, in another species, cross less readily, if at all. Some antibiotics, antifungal medications, and antiviral medications pass through the placenta to create birth defects in animals, but not humans.[62] Contaminants cross the placenta more frequently than most imagine, but happily not usually in quantities sufficient to produce birth defects. For example, one study found PCBs (polychlorinated biphenyls) and DDT (dichlorodiphenyltrichloroethane) in thirty percent of all women tested during amniocentesis.[63]

So, not all species are equally susceptible to teratogenic influences by any given chemical. Likewise, an agent that is teratogenic in some species may have little or no teratogenic effect in others.[64] Of over 1,200 chemicals found to cause birth defects in animals, only thirty cause them in humans, according to the *New England Journal of Medicine*.[65] How many of the 1,170 chemicals that are benign in humans became valuable drugs despite animal testing or how many could become valuable drugs if they could bypass the animal assays? This conundrum is not new. Twenty years ago, Dr. Schmid stated:

> More than eight hundred chemicals have been defined as teratogens in laboratory animals, but only a few of these, approximately

twenty, have been shown to be teratogenic in humans. This discrepancy can be attributed to differences in metabolism, sensitivity and
exposure time.[66]

Articles in many other publications repeat these conclusions.[67,68,69,70]
Twenty out of eight hundred is only 2.5 percent effectiveness. Not a
number on which we would want the life of our child to depend.

We do not deny that after epidemiology or clinical observation links
a drug to birth defects, as it did thalidomide, animals can usually, though
not always, be found to demonstrate that effect. Just as Karnofsky
postulated, if researchers try hard enough they may eventually inflict
birth defects on some animal species with a substance that is teratogenic
in humans. But to what purpose? Animal experiments that are not
predictive are of no value. They just use up money and other valuable
resources. There is no sense in "validating" something that human data
has already rendered conclusive. The whole point in using animal models
is to avoid birth defects. If the models do not predict accurately and
reliably which drugs are dangerous to the human fetus and which drugs
a pregnant woman can consume if she needs them, they are poor models.

When a drug tests "positive" for problem-causing in animals, but later
proves healthful to humans, that is a *false positive*. When a drug tests
negative for problem-causing in animals, but later proves deleterious to
humans, that is a *false negative*. Chemicals that are teratogenic in animal
tests may or may not indicate danger to humans, and negative results
may incorrectly imply the absence of risk. The bottom line is, therefore,
that "final proof of whether a chemical is likely or not likely to be
teratogenic in man can only be sought in man."[71] As Dr. Ralph Heywood wrote in 1978:

> There is fundamental lack of knowledge of pathogenesis of most
> malformations and, therefore extrapolation to man has to be done
> with reservation and care. Negative results cannot be used to predict
> that an agent will lack teratogenic effect in man. . . . Surprisingly
> few compounds have been shown to be teratogenic in man although
> a large number, including aspirin, steroids, vitamin A and B, insulin
> and hydantoin, have been shown teratogenic in rats.[72]

There are innumerable examples of false positives, medications once
thought to be teratogenic that have now been shown not to be. Perfectly
human-safe drugs can unleash a veritable sideshow of deformities in
rats, mice and rabbits. Of these, aspirin is the most common example.
Yes, our most popular drug, grandfathered in before the requirement for
animal validation was legislated, would never have made it through the
animal assays now required. Aspirin produces birth defects in mice, rats,
guinea pigs, rabbits, cats, primates, and dogs. Yet, eighty percent of all

pregnant women take aspirin and do so without a problem.[73,74] Even this one incidence shows how absurdity of testing animals for human teratogenesis. And aspirin is the tip of the iceberg. Dextromethorphan, a medication used to treat coughs, is one of the latest medications initially thought to cause birth defects because of animal testing; it appears to be entirely safe in humans. In typical fashion, the media highly publicized a report of neural tube defects in chick embryos that were exposed to dextromethorphan. One of the physicians who authored the article, a Dr. Koren, stated that animal studies do not always predict teratogenicity in humans.[75]

The following list indicates just a smattering of other now-valuable human drugs that proved teratogenic in typical lab animals: The diuretics dichlorphenamide, acetazolamide, ethoxzolamide, furosemide, methazolamide, and spironolactone all produced birth defects in rats, mice, rabbits and hamsters and various other animals.[76] The antihypertensive medications diazoxide, hydralazine, reserpine and guanabenz produced birth defects in sheep, mice, rabbits, and rats. The analgesics aspirin, acetaminophen, codeine, hydrocodone, hydromorphone, meperidine (Demerol), morphine, oxymorphone, phenazocine, and propoxyphene produced birth defects in numerous species.[77] The antigout medications colchicine and allopurinol did likewise.[78] Spermicides, benzodiazepines, antibiotics like streptomycin and penicillin, antihyperglycemics like insulin and ten noninsulin antihyperglycemics all produced birth defects in some animals.[79] Anesthetics such as enflurane, ether, halothane, Isoflurane, methoxyflurane, nitrous oxide, Sevoflurane, and procaine were teratogenic in various species.[80] Other teratogenic drugs are as follows: corticosteroids, allergy and respiratory medications,[81] antinausea medications,[82] antimalarials, antiparasitics, anthelmintics,[83] antibacterials such as sulfa drugs, ampicillin, cephalothin, chloramphenicol, erythromycin,[84] thyroid disorder medications,[85] nonsteroidal anti-inflammatories, immunosuppressants, anticoagulants such as warfarin, antineoplastics, angiotensin-converting enzyme inhibitors, aminoglycosides,[86,87] and calcium channel blockers.[88]

Dr. Dennis Parke stated in 1983:

> corticosteroids are known to be teratogenic in rodents, the significance to man has never been fully understood, but nevertheless is assumed to be negligible. However, the practice of evaluating corticosteroid drugs in rodents still continues, and drugs which exhibit high levels of teratogenesis in rodents at doses similar to the human therapeutic doses are marketed.[89]

Corticosteroids, like cortisone, produce malformations in almost all animals tested, but not humans.[90] Another case in point is Bendectin, an antinausea medication used by up to forty percent of all pregnant women in the United States. Animal studies suggested that the drug was teratogenic; it was withdrawn but later shown to be safe. Believers in the efficacy of the animal model should be asking themselves this: If these dozens of drugs cause malformation in animals, how did they eventually make it to market? They made it to market because those in charge know animal testing is done purely as a liability issue, not a scientific issue. The animal testing continues because it is profitable to human researchers, though not for human patients.

What about false negatives? Sometimes the FDA approves drugs because they show no ill effect in animals, then later human birth defects develop. In 1998, the *New England Journal of Medicine* published an article on birth defects and medications.[91] The article pointed out that one-half of all pregnancies in the United States are unplanned, and hence any medications being taken by women of childbearing ability could theoretically be taken by a woman who is unknowingly pregnant. Tragically, these women sometimes miscarry or give birth to babies with birth defects. Sadly enough, these circumstances provide our best indicators of teratogenic medications, false negatives.

Whereas animal models mislead, epidemiology continues to find medications that were tested on animals but proved harmful to the fetus. It was epidemiological data (as well as clinical experience) that ascertained that the aminoglycoside antibiotics streptomycin and kanamycin led to eighth cranial nerve damage in the fetus and that tetracycline resulted in discolored teeth and poor bone growth. This was not evident in animal models. Population studies led to the knowledge that antiepileptic drugs—especially carbamazepine, valproate monotherapies or polytherapies—taken during pregnancy can be teratogenic. They increase the risk of congenital abnormalities like congenital heart defects, hypospadias, polydactyl, spina bifida aperta, and facial clefts.[92]

Among the drugs that cleared animal assays and later caused teratogenicity in humans are methotrexate, carbamazepine, valproic acid, retinoids, lithium, and danazol. We learned about the effects of DES (a hormone replacement therapy) through epidemiology not animal studies.[93,94,95] Accutane, an acne medication, is so profoundly teratogenic in women that each package comes with a label showing a pregnant woman with a red-barred circle on her. Since the most likely target patients for an acne medication tend to be women in their childbearing years, it was discussed that the drug should be restricted from being prescribed for women at all. This is one of the retinoids, which made it through animal teratogen testing without problems. When clinical observation showed that the very popular warfarin, a blood thinner more commonly known as Coumadin, caused teratogenicity, researchers went to work on animals, but were then only able to reproduce the effect in rats and not any other animals.[96] The ace-inhibitors, captopril and enalapril, are strongly suspected of causing birth defects in humans but not animals.[97]

In 1984, Dr. Lasagna stated:

> False positives and false negatives abound. Once one has established that a drug is a teratogen for man, it is usually possible to find, retrospectively, a suitable model. But trying to predict human toxicity—which is after all what the screening game is all about—is quite another matter. Cortisone is a potent dysmorphogen in the rabbit and mouse, but does not produce malformations in the rat. Carbutamide produces malformations in the eyes of rats and mice, but facial and visceral malformations in rabbits.[98]

Of this same syndrome, Dr. George Lin wrote in the journal *In vitro Toxicology* that

> . . . there is no ideal animal model to extrapolate teratogenicity results to human exposure because of species sensitivity and species difference.[99]

Lin's concern is Lasagna's concern—that of a suitable model. It is exactly the story of this book. Despite immense labor and many millions of dollars, researchers have not come up with a single species whose system adequately mirrors that of a human. Rats are widely touted teratogenic pseudo-humans, and supporters of animal models claim that rats get birth defects from every chemical that causes birth defects in humans. But this claim is meaningless for three reasons. In high enough dosage, of course rats will produce malformed offspring. Many medications, after due diligence with animals, usually rats, have been released on the unsuspecting public only to be later shown to harm the fetus. Also, many now valuable human drugs cause teratogenesis in rats. The comparison is not an honest one.

In 1980, Dr. S. J. Yaffe stated:

> Experiments with animals have yielded considerable information concerning the teratogenic effects of drugs. Unfortunately, *these experimental findings can not be extrapolated from species to species or even from strain to strain within the same species, much less from animals to humans* [emphasis added].[100]

In the book, *Monitoring for Drug Safety*, Dr. Smithells states:

> In the absence of useful tests for teratogenicity, clinicians [physicians] have to accept the responsibility for drug exposure in early pregnancy. . . . If clinicians were more aware of the shortcomings of

animal teratogenicity testing, they might take this responsibility more seriously.[101]

That is to say that for physicians, the only common sense approach to prescribing medications to a pregnant patient is to do so only if absolutely necessary. Although this is somewhat counter to our "pill for every ill" mentality, it would safeguard many unborn babies from potentially dangerous medications. The physician and the mother must ask what the risk benefit ratio is before prescribing or taking a medication. Hence, patients who are pregnant or intend to become pregnant should ask for *human data* in regard to the teratogenic impact of any drug they intend to take. They should make certain, with their doctors, that the drug is necessary. This same question should really be asked before *anyone* takes any medication.

Birth Defects and Childhood Illnesses

By supplying more data about congenital disease, clinical observation, and epidemiology have in some cases had preventive consequences. By knowing more about the influences that might cause these diseases, parents can better plan their childbearing. And, when adverse exposures are potentially part of the cause, they can act on this knowledge to avoid these exposures.

Down's Syndrome

Down's syndrome, which used to be called *mongolism,* has been described since antiquity. Thanks to *in vitro* studies, we now know that Down's syndrome children are born with an extra chromosome.[102] With severe to moderate mental retardation, their IQ ranges from twenty to sixty; they may have heart defects, leukemia, and other ailments that decrease their life expectancy. With the advent of chromosomal analysis, amniocentesis, and epidemiology, we have been able to quantify risk and learn more about why the disease occurs. It is evident that, for instance, older women are more likely to give birth to an affected infant than are younger mothers. Just over twenty percent of Down's syndrome infants are born to mothers over thirty-five years old.[103] The number born to twenty-five-year old mothers is approximately one in two thousand, but the risk increases to one in two hundred for thirty-five-year olds and more than one in forty for women older than forty years. The epidemiological proof that older mothers are more likely to have a child with Down's syndrome has the effect of pressing families to plan parenthood earlier. Here again, there is an imperative for prenatal care. Tests such as ultrasound measurement of fetal nuchal-translucency, amniocentesis, and chorionic villus sampling can be used to detect the chromosomal abnormality causing Down's syndrome. In addition, maternal blood tests suggest the presence of a fetus with Down's syndrome.

Congenital Diaphragmatic Hernia

Congenital diaphragmatic hernia (CDH) is present in one of every 2,500 neonates. CDH-affected babies have healthy hearts that can pump blood, but their lungs are not able to oxygenate blood because a hole in the diaphragm allows the contents of the abdomen to compress one or both lungs. This is potentially life threatening. The smaller amount of abdominal contents in the chest, the more the lungs are able to develop. Children born with a less severe form are more likely to live than those with all the abdominal contents in the chest. Sometimes immediate surgery can position the abdominal contents back in the abdomen and seal the hole, which allows the lung to expand.

If the compressed lung has not developed normally due to the compromised blood flow *in utero*, urgent surgery will not correct the problem. Technology has improved survival rates for these babies in several ways, none of them attributable to the animal model, though the Foundation for Biomedical Research would have you believe otherwise. It claims that one CDH treatment, extracorporeal membrane oxygenation (ECMO), resulted from animal models. ECMO replaces the lungs. The principle is quite similar to the cardiopulmonary bypass for heart surgery. As discussed in the chapter on coronary artery disease in *Sacred Cows and Golden Geese,* animal experimentation actually misled the development of the cardiopulmonary bypass. When animal data was employed in developing the ECMO, the bypass catastrophe repeated. Only by substituting knowledge from blood-flow studies based on humans for the animal data did the technique work. Additionally, the use of a pulse-oximeter, a device that looks like a piece of tape on a cord, allows clinicians to monitor—noninvasively—how well the baby is doing by telling them how well the tissues are being oxygenated.

Autopsy and ultrasound deepened the understanding of CDH and allowed doctors to diagnose the defect in fetuses. Marvelously, doctors eventually learned to operate on CDH-affected babies before birth. Repairing the diaphragmatic defect *in utero* allows the lungs to develop normally, thus obviating the need for ECMO. The March of Dimes credits animal studies, which it funded, for the development of this surgery. However, a look at history again disproves this claim. Researchers at Johns Hopkins University did learn that a diaphragmatic hernia can be simulated in unborn lambs by putting catheters in lambs' chests. However, from this procedure they gleaned nothing new concerning prevention or cure. In an effort to gear up for pre-birth surgeries on children, doctors practiced emptying the abdominal contents out of the chests of these embryonic sheep, employing a technique that surgeons had used to repair the defect in babies after birth. Despite practice on sheep, five out of the first six human babies died. One reason for this was differences between sheep and humans. Since the artificially induced hernias in sheep were not the same as CDH in humans, the lamb surgeries could not predict complications in humans. It was essentially a different operation. Surgeons admitted that the first failed attempts in babies, not sheep experiments, taught them how to perform the operation in humans.[104] Dr. Ferdinando de Leo, professor of Pathological and Clinical Surgery at the University of Naples, states:

> I have been a surgeon for 51 years. I am still performing operations daily, and can state that in no way whatever do I owe my dexterity to animal experimentation. . . . If I had had to learn surgery through animal experiments I would have been an incompetent in this field, just as I consider those of my colleagues to be incompetent who say that they have learned surgery through animal experimentation. It's true that there are always advocates of vivisection who say that one must first practice on animals in order to become a surgeon. That is a dishonest statement, made by people who reap financial benefit from it.[105]

Erythroblastosis Fetalis

Erythroblastosis fetalis is a hemolytic disease of the fetus and newborn. It has in all probability killed babies throughout history, but it was first described in 1609 in a set of twins. As the condition destroys red blood cells, the babies usually have severe physical abnormalities and often die. Blood type incompatibility with the mother is the cause of erythroblastosis fetalis. It occurs when the baby is RhD+ and the mother is RhD− but has been exposed to RhD+ through another pregnancy. Prior to 1998, amniocentesis was the only way to diagnose the problem. Since then, *in vitro* researchers devised an alternative. A blood sample from the mother will tell physicians if the baby is at risk for the disease and if intrauterine transfusions should be implemented.[106]

Esophageal Atresia

Esophageal atresia is a condition affecting roughly one in four thousand babies. A gap between the esophagus and the stomach interferes with feeding. Surgical procedures developed in the 1940s can correct the condition, but advances in parenteral nutrition (feeding by a means other than the mouth) and advances in the ability to ventilate the lungs artificially have also helped these children survive. Surgeons can treat other abnormalities of this nature, such as imperforate anus and necrotizing enterocolitis, as a result of technical innovations and advances in clinical medicine.

Reye's Syndrome

Sometimes, children who received aspirin or aspirin-like medications for otherwise unremarkable viral illnesses, such as chicken pox or influenza, manifest Reye's syndrome. Their brains and livers swelling, afflicted children may come down with fever, severe vomiting, confusion, and sometimes, respiratory arrest. Reye's syndrome can lead to coma and other serious or life threatening conditions, and was during the 1970s and 1980s feared as the next polio. It was named after Australian physician Ralph Douglas Kenneth Reye, whose epidemiological data dispelled this fear by confirming the association between taking aspirin-like medications during a viral illness and the risk of developing this dangerous syndrome.

Since aspirin was making money, it was difficult to persuade those with a vested interest that its use should be, in some cases, curtailed. Researchers conducted many unnecessary studies attempting to show that it was not deleterious. After disapprobation of the administration of aspirin-like medications during certain illnesses by the medical community, the incidence of Reye's syndrome fell dramatically. This is another example of an ounce of prevention being worth more than a million pounds of animal experimentation. Today, Reye's syndrome affects very few children, thanks to prevention brought about by epidemiology.[107,108]

Cytomegalovirus

In the early part of the twentieth century, pathologists noted that certain cells contained odd inclusions in children presumably dying of syphilis. In 1921, two scientists Goodpasture and Talbot guessed that the inclusions were a virus. In 1956, using *in vitro* research, the virus was isolated, and epidemiological studies better clarified the disease process, now called cytomegalovirus (CMV). Mothers exposed to CMV through other children may infect infants both prenatally and at birth. CMV brings on mental retardation, liver and spleen problems, respiratory distress, and developmental abnormalities in infants. Up to one-third die.

Sudden Infant Death Syndrome

Sudden infant death syndrome (SIDS) is the unexpected and sudden death of an apparently healthy infant. The leading cause of postneonatal infant mortality in our country, SIDS is a mass apprehension that many animal experimenters have not failed to exploit. SIDS does not appear to be a specific disease or condition, but rather the result of a variety of factors that may rob an infant of his or her life. Among those factors are pre-term birth, low-birth weight, young maternal age, late or no prenatal care, smoking and substance abuse in the mother, and postnatal exposure to cigarette smoke. Recently, clinical studies confirmed that babies born to mothers who smoke during pregnancy or who are born into a household of smokers have an increased risk of SIDS. Sharing a bed with a mother who smokes and sleeping in a bed with loose bedding capable of covering the baby's head are documented contributors.[109,110,111,112] In some cases, placental abnormalities during pregnancy disrupt the normal delivery of oxygen and nutrients.[113] Long-QT syndrome may also play a role. (QT is an interval of the heartbeat, as monitored on an EKG tracing. The heartbeat—electrically speaking—is referred to as a P-Q-R-S-T complex. P is when the atria contract. QRS is the ventricular contraction. The QT interval is the length of time between the beginning of ventricular contraction and repolarization of the ventricles.) The mutation arises spontaneously—that is, it is not inherited from the parents. In 1998, researchers in Italy collected data on 34,000 infants who underwent electrocardiography. They found that that fifty percent of infants who died of SIDS had a prolonged QT interval.[114,115]

Observation of human babies disclosed the relationship between SIDS and babies sleeping on their stomachs. The incidence of SIDS has declined by thirty percent since the Centers for Disease Control began recommending that infants should not sleep on their stomachs but on their backs or sides instead.[116] It is down by two thousand babies per year.[117] Since we are one of the few animal species who routinely sleeps on its back, this would be a hard thing to reproduce in an animal.

Miscellaneous Congenital Defects

The roster of wisdom recovered from epidemiology goes on and on. Epidemiology supplied very useful information about fetal hydantoin syndrome. As early as 1941, epidemiological evidence in Australia established the connection between maternal rubella and birth defects. The incidence of cataracts in newborns rose after a rubella outbreak.[118] Epidemiology revealed that the risk of retinopathy of prematurity (retrolental fibroplasia), which causes blindness in newborns, can be diminished by giving only as little oxygen as possible for as short period of time as possible to premature babies. This does not always prevent blindness but it does lessen the risk.[119]

As an example of scientific excellence in pediatrics, take Hudson Freeze, of the Burnham Institute in La Jolla, California. Dr. Freeze observed that cells from the mold *Dictyostelium* were similar to the cells from children suffering from a rare disease known as congenital disorders of glycosylation type1b (CDG1b). Patients lack a fundamental enzyme called phosphomannose isomerase that the body employs in the manufacture of numerous vital compounds. Disorders of the blood produce an array of suffering and the children can bleed to death.

The mold had been engineered to lack the same enzyme. When Freeze added a sugar called *mannose* to the mold cells, they produced the compounds that previously they could not. He then did the same thing to cells taken from children with CDG1b and found that those cells were now producing the compounds the body needs. Freeze published a paper about his discovery that was read by a physician in Germany treating a patient with CDG1b. The boy was dying. His physician contacted Freeze who told him how to give the mannose and the boy recovered. This scenario reveals how many medical breakthroughs occur. First serendipity plays a part, and then a real-life crisis necessitates the use of the discovery.[120]

Inherited Diseases

Sickle Cell Anemia

Sickle cell disease or sickle cell anemia (SCA), a disease of the hemoglobin, is an inherited disease. First described in 1920, it is the single most common gene disorder in black Americans, affecting one out of every 375 people of African ancestry. Our understanding of sickle cell anemia has accumulated through clinical observation and *in vitro* research. *In vitro* research revealed the genetic mutation that leads to the disease—a single change in the DNA base sequence at a particular gene, also called a single nucleotide polymorphism or SNP, which we discussed in the Internal Medicine chapter. At position 6 in a sequence that includes 146 amino acids, the amino acid valine is substituted for glutamic acid. This alters the structure of the hemoglobin protein, which this particular gene directs. With this substitution, the defective gene cannot produce normal red blood cells.

In victims, hemoglobin molecules deform, and red blood cells lose the flexibility to pass through blood vessels. The cells form a sickle shape. Red blood cells clog blood vessels similar to the way cholesterol clogs arteries; therefore, parts of the body are deprived of oxygen. When body tissues become *hypoxic* (oxygen-deprived), it causes severe pain. The body is forced to destroy red blood cells as they choke the vessels, and anemia ensues. The abdomen and heart may enlarge, and extremities swell painfully. Disturbances in blood flow also dispose affected persons to infections.

Affected children suffer what is referred to as *sickle crisis*. This is a very agonizing event. Only powerful narcotics in combination with other therapies can decrease the pain level, and the crises eventually require more and more medication. Even the strongest medications may be unable to mitigate these excruciating episodes. Stroke, heart attack, and other infarcts are common.

Sickling of red blood cells occurs in some animals too, but the pathological symptoms seen in humans are unique to humans. Deer sickle, but by an entirely different mechanism. The affliction also appears in the mongoose, sheep, goats, hamster, squirrel, and others.[121] Researchers have made a concerted effort to get mice to sickle. They tried inserting human genes into mice, but that did not reproduce the effects of the disease.[122,123,124,125,126,127] Next, researchers specifically implanted the gene responsible for sickle cell anemia into mice, but still the progeny mice did not exhibit sickling.[128]

Eventually, scientists did induce sickling in a particular strain of mice, but the lengths to which they had to go invalidate any results for humans. Basically, they took mice with another hemoglobin disorder, *thallasemia,* and mated them with mice in which they had induced a variant of human sickle cell anemia. This enterprise created a whole new disease in mice, with superficial similarities to sickle cell anemia and minor sickling. The minor sickling was insignificant.[129,130] Trying to correlate the slight sickling in the mice and the drastic sickling in humans is a huge stretch. The malconformation is like drawing parallels between the common cold and AIDS because they are both caused by viruses.[131,132,133] Numerous and repeated attempts have not reproduced the human disease in mice, because, as one researcher said, "The transgenic mouse [as any other animal model] will necessarily differ substantially from the sickle cell disease patient because mouse anatomy and physiology differ from human."

Even the highly touted knockout mouse—with human hemoglobin genes—does not express sickle cell anemia. Dr. Ronald Nagel, who works for an institution that profits from animal experimentation, admits that "there is still no perfect model of sickle cell anemia—indeed the perfect model may be impossible."[134]

Test-tube studies of blood from human patients uncovered that fetal hemoglobin had less sickling than did the adult form.[135,136,137,138] Researchers have attempted to get infant baboons to sickle, without success. As infants, baboons make "fetal hemoglobin," as do humans. Researchers successfully induced baboons to make more fetal hemoglobin. One of the medications used in this effort, 5-azacytidine, was then given to human sickle cell anemia patients. It was not successful in humans.[139] Baby baboons also received erythropoietin to treat the induced version of the disease, but again humans did not respond the same way.[140] In summary, though many animals experience sickling of their red blood cells, none do so the way humans do.[141] Some scientists have stated that:

The mechanisms involved and very probably the molecular pathology in these animal models differs considerably from that postulated [for sickle cell anemia].[142]

Meanwhile, advances in SCA have come from the fields of molecular biology, hematology, pathology, and clinical observation.[143] It is hoped that the ability to spot the disease *in utero* may lead to new treatment methods. Physicians can now screen unborn babies for SCA as well as diagnose newborns via the *in vitro* tests of DNA analysis, hemoglobin electrophoresis, isoelectric focusing, or high performance liquid chromatography. Couples who carry the sickle cell trait can have the eggs fertilized and tested for sickle cell prior to implantation in the uterus. The test employs sophisticated *in vitro* technology.[144]

Scientists have used human studies to create one method of increasing fetal hemoglobin in humans. By testing blood from diabetic patients and newborn infants of diabetic mothers, they isolated a chemical that increases human fetal hemoglobin.[145] This chemical, butyrate, has not yet undergone clinical trials. Injecting normal red blood cells into fetuses that have the gene for sickle cell anemia is another promising treatment that may preclude the disease.[146] Clinical studies combined with ultrasound evaluation showed that transfusions in young patients suffering from sickle cell anemia prevented strokes that would have occurred before they reached maturity.[147]

Since 1975, the mortality and morbidity associated with SCA has decreased. Antibiotic administration, vaccines, proper nutrition, and routine physical checkups somewhat keep the risk of infection in check. Transfusion therapy and hydroxyurea are also helpful.

Cystic Fibrosis

Cystic fibrosis (CF) usually appears in infancy. Like sickle cell anemia, CF is an inherited disease attributed to a single nucleotide polymorphism. One out of every twenty-five children of northern European descent carries the gene and one out of 2,500 is born with the disease. It affects as many as 30,000 people in the United States alone. The gene defect disallows chloride ions across cell membranes and creates an imbalance of fatty acids. In victims, levels of docosahexaenoic acid are low and those of arachidonic acid are high. As a result, thick secretions plug the lungs making breathing difficult and increasing vulnerability to respiratory infections. The disease also interferes with the digestive tract. The pancreas acquires cysts and fibrous degeneration, impairing the ability to absorb nutrients. The lungs and pancreas begin to fail. Autopsy, not animal experimentation, suggested this etiology. Pancreatic cysts were found routinely in affected individuals. Further studies exposed changes in the lungs and intestinal ducts.[148,149,150] Children with cystic fibrosis risk dying if overheated. Even with early diagnosis and specialized care, a mere half makes it to young adulthood. Only organ transplants can ultimately forestall death.

Nonanimal methods have led to tremendous progress in understanding CF. Observation of patients whose perspiration was very salty first identified cystic fibrosis. This clinical experience led to the "sweat test," which is still used today.[151] During a heat wave in the 1940s, physicians caring for children suffering from heat prostration noticed that many were affected with CF. They had very low levels of sodium in the lungs, however. Studying the chloride concentration in the urine and sweat uncovered the involvement of the chloride ion in interrupting salt transport.[152,153,154] In the 1980s, *in vitro* research clarified the exact mechanism of defective chloride transport in cystic fibrosis patients.[155,156]

In 1989, scientists located the genetic cause of cystic fibrosis using *in vitro* technology. A series of 1,480 DNA triplets (4,440 base nucleotides) code for the correct protein in unaffected people. In patients suffering from cystic fibrosis, only one of the triplets is different—a minute difference which leads to very major consequences. Over ninety-nine percent of the genetic material is identical but the fraction that is not—one out of 1,480—shortens life and causes great suffering. This same gene, the CFTR gene also plays a role in non-CF patients suffering from pancreatitis.[157,158] It appears to control chloride channel function and chloride transport.[159] From here, emerged the best diagnostic test for cystic fibrosis, DNA analysis, and in fact, cystic fibrosis can now be diagnosed *in utero*.[160]

Next, scientists found that swabbing the inside of the nose or collecting lung washings from human patients supplied affected cells. These cells grow continuously in the lab, affording an unlimited quantity of test material, but that did not impede further animal experiments. Scientists can insert the proper human gene sequence *in vitro*, correcting the damaged portion of the gene responsible for cystic fibrosis. However, instead of inserting the correct DNA sequence into humans with cystic fibrosis to begin with, and despite the general consensus that there is no "animal model" for cystic fibrosis, researchers hastened first to induce cystic fibrosis in animals by creating transgenic mice. After, they performed the same insertion.[161,162,163,164] In an article in *Hospital Update* January 1994, researchers stated:

> Mouse bronchial epithelium is quite different from human; in particular it does not contain serous glands, which express the majority of CTFR in human beings, so care will need to be exercised in extrapolating any findings to man.[165]

The animals do not exhibit the same changes in the pancreas that humans have. Neither do they face the lung infections that cystic fibrosis patients do.[166] Considering the fact that the pancreas and lungs are the main organs affected, one can say that the transgenic model of cystic fibrosis is a failure. Moreover, the mice died of intestinal disorders before the lung abnormalities manifested. This was fruitless experimentation. It was impossible to ascertain whether the gene prevented anything since animals do not get cystic fibrosis.[167,168,169,170] Authors of a previously cited article wrote in this regard:

> It is apparent from an analysis of some transgenic disease models that the actual benefits of using the models are rarely completely equivalent to the potential benefits. . . . The currently available transgenic models for cystic fibrosis (CF) illustrate this point. None of the strains is ideal, with either the genotype and/or the phenotype of the mouse failing to accurately model the human condition.[171]

Meanwhile, real patients continued to suffer and die. Finally, researchers administered the gene therapy to human patients. It appears to be a safe, effective method of improving chloride transport, but the method is far from perfect and more human studies need to be done.[172]

The effect of CF on the pancreas diminishes the availability of pancreatic enzymes required for digestion. Therefore, CF patients must take those enzymes in pill form. Most of these come from pigs as a byproduct of slaughterhouses, just as insulin did at one time. However, they can and probably soon will be exclusively manufactured from chemicals, plants or fungi, thus eliminating any infectious-disease risk.

Scientists now know that defective proteins impede the flow of salt and water through microscopic channels in cells within the lungs and other organs. This is what leads to the sticky, thick secretions, setting the stage for life-threatening lung infections, the leading cause of death in CF patients.[173] Unlike unaffected people, their lungs do not kill inhaled and aspirated bacteria to an adequate extent. They are less equipped to handle lung infections because the natural antibiotic substances that normally line the lungs are too salty to be effective. Researchers felt that coating the lungs with a simple sugar, Xylitol, might lower the salt concentration in the liquid that covers cells lining the inside of the lungs, thereby enhancing the bacteria-killing activity of the body's natural antibiotics. This proved a viable treatment in human volunteers.[174]

Some cystic fibrosis therapies focus on thinning very thick secretions that the lungs have been unable to clear. Aerosolized medications are among the popular techniques. The idea to place secretion-loosening medications in an inhalant spray came from clinical observation of tuberculosis patients in the 1940s. Physicians noticed that tuberculosis patients given aerosolized treatments could cough up their secretions more easily than the patients not given these treatments. Physicians then applied this observation to other patients with lung problems, such as cystic fibrosis. Many years later, researchers experimented on rats to see if the chemical used to aerosolize the medications was toxic to the lungs. They found that it was not (no surprise since it had been used for years) and reported that their findings were "in complete agreement with the [human] clinical experience."[175,176] The rat studies unveiled nothing new.

The success of early aerosolized treatments opened the field to look for other medications that could be aerosolized to thin the secretions. Scientists conducted test tube research to find medications capable of thinning the secretions. They did this by collecting secretions from cystic fibrosis patients and combining them in a test tube with various medications. They found that bovine pancreatic DNase I was very effective in thinning the secretions. Clinical trials were consistent with this finding. It worked well, however some patients did exhibit adverse reactions, probably to the bovine protein or a contaminant in it. This problem has been dealt with by replacing the bovine DNase with human DNase (rhDNase). So, once again the human version worked better for humans than the animal version.[177,178,179,180,181]

In any event, rats make entirely inadequate pseudohumans for diseases of the respiratory system such as cystic fibrosis, as reinforced by this quote from the *Handbook of Laboratory Animal Science*:

> One of the most commonly used species in toxicology research, the rat, differs substantially from humans: It lacks a gall bladder, is a very effective biliary excretor, displays less efficient plasma binding of drugs, is an obligate nose breather, is nocturnal, and has a different teratogenicity, to name only a few dissimilarities. Rats are consequently considered to be inadequate predictive models for, e.g., human asthma or bronchitis research, but have nevertheless been extensively used for the experimental study of bronchitis.[182]

Mortality rates from CF decreased with the advent of physical therapy, antibiotic treatments, and medications originally used for high blood pressure and other diseases.[183,184] These treatments came about because of nonanimal research modalities.

The study of asthma has not fared any better using animals for the same reasons. Scientists state that:

> . . . at present, there is no ideal animal model of chronic asthma and such a seemingly complex disorder may be impossible to mimic.[185]

And:

> Animal models have fallen short of reproducing the human disease, particularly in mimicking the spontaneous and persistent airflow obstruction that characterizes asthma.[186]

Ears, Hearing Impairment, and Deafness

The most common inherited sensory disorder is hearing impairment or deafness, afflicting approximately one child in a thousand. Most patients, deaf from birth, have abnormalities in the outer or middle ear. Human epidemiology studies and *in vitro* research techniques recently identified genes responsible for deafness in the most common manifestations, Pendred's syndrome and non-syndrome deafness.[187] After ascertaining families with higher than expected incidence of congenital deafness, researchers drew blood samples, or obtained skin samples, and analyzed them for chromosomal and gene abnormalities. By studying the DNA, researchers were able to isolate which genes are responsible.[188] In one case, researchers studied the records of an extended family dating back to 1713, examining the genes of 147 family members. They found that the gene DFNA1 was the cause of deafness.[189] Deafness, both congenital and sporadic, has been linked via epidemiology and *in vitro* research to the gene for connexin-26 on chromosome number 13. Investigators recommend testing hearing-impaired newborns for this mutation because they may require aggressive intervention and it will eliminate the inquiry for other diseases.[190,191] Other epidemiologic and *in vitro* studies have found more genes that may be responsible for deafness and hearing related disorders.[192] A hearing aid can sometimes treat hearing impairment, either congenital or acquired. The hearing aid was invented in 1880 without animal testing.

Despite this, we are still doling out money for deafness research on animals. For example, congenital deafness is prevalent in Dalmatian dogs, and Dalmatians that have fewer spots are higher risk. However, we are still waiting to understand how that applies to humans. White cats with blue eyes are also at increased risk for deafness—one more fact not applicable to humans. Another hearing related problem is "tinnitus" or ringing in the ears. Although cat research suggested that tinnitus initiated in the cochlea (part of the inner ear), PET scanning determined that tinnitus begins in the auditory center of the brain. Further research will no doubt cast more light but it appears that animal experimentation once again misled researchers.[193] Genetics may also be to blame for susceptibility to some earaches. A human study indicated a genetic component in frequent otitis media and middle ear effusion.[194]

Other Genetic Diseases

Again and again, a combination of epidemiology and *in vitro* research has defined genetic diseases. Epidemiologists first remark on similar conditions among many. Then geneticists put cells from the afflicted under scrutiny and find out why. Sometimes clinical observation plays a role.

In 1858, Dr. Duchenne described muscular dystrophy, a disabling disease that gradually wastes away skeletal muscle. The advent of molecular biology made it possible to observe the chromosomal disorder that causes the disease, which though carried by females, almost always affects males.[195] Muscular dystrophy stems from a gene mutation that prevents the body from making dystrophin, a protein that allows muscles to contract. Using gene transfer therapy to replace the defective gene has long been a dream but may be closer to becoming a reality. The problem with adding the healthy gene has been that the gene was much larger than the virus vector used to get the gene inside the cells, hence it could not be transferred into the cell. Dr. Xiao Xiao, a researcher at the University of Pittsburgh, used *in vitro* research to design a smaller gene. He took the parts of the gene coding for dystrophin that were most important and discarded the rest. He followed with *in vitro* research to see if the outcome was viable. It was. He then performed the procedure on animals. That part revealed nothing new. It had already been shown to work *in vitro*. The next step is to try it in humans.

Genetics has allowed us to find the basis of many other diseases. A defect in chromosomes 8 and 20 can lead to congenital neonatal epilepsy. Prader-Willi syndrome has been traced to a deletion of the paternal chromosome 15. Giant axonal neuropathy stems from abnormalities on chromosome 16.[196] Human research determined the gene that results in Peutz-Jeghers syndrome. Peutz-Jeghers syndrome causes cancer and other diseases.[197,198]

Dr. Andreas Rett identified a childhood neurodevelopmental disorder characterized by normal early development followed by loss of purposeful use of the hands, distinctive hand movements, slowed brain and head growth, gait abnormalities, seizures, and mental retardation. Rett syndrome, as it is called, affects females almost exclusively. Geneticists found mutations in an X-linked gene, MECP2, that seem to play a role. Scientists involved are hopeful that recognition of the gene will lead to treatments that can be administered early, before symptoms appear.[199]

Genetic research found a defect on a mitochondrial gene that causes blindness in young adults, Leber's hereditary optic neuropathy. Finding the cause of human optic neuropathy was an important event. Only with epidemiological studies and *in vitro* research was human disease elucidated. Mitochondrial genes differ from the more familiar genes on the chromosomes. All cells contain mitochondria; the powerhouses of the cell, and within each mitochondrion are loops of DNA, each containing thirty-seven genes. Maternal mitochondrial genes are passed along to offspring; therefore babies will suffer from their mother's mutated mitochondrial genes. No paternal mitochondrial genes are passed on. Progressive muscle disorders, premature aging, some forms of heart disease and diabetes, and possibly diseases such as Alzheimer's, have also been shown to be caused by flawed mitochondria.[200]

The difficult-to-diagnose Familial Mediterranean Fever (FMF) is another hereditary disease that condemns its victims to suffering, lifestyle changes, and death. Epidemiology suggested that FMF was hereditary in people of the Near East such as Arabs, Sephardic Jews, Armenians, and Turks. Subsequent *in vitro* research and epidemiology allowed scientists to isolate and clone the gene.[201] This knowledge will allow physicians to diagnosis FMF unequivocally and may allow fetal testing.

Williams' syndrome is a congenital defect in which children, whose facial features include a small chin and turned up nose, have low intelligence quotients but are gifted in music and language. Williams' syndrome children suffer from medical problems throughout infancy and childhood. Clinical observation by a heart specialist in 1961 revealed that some children with a particular heart problem also demonstrate mild mental retardation and the aptitudes just described. Their typical "pixie" image completed the physical picture. Epidemiology, clinical research, and *in vitro* research led the identification of the gene responsible for Williams' syndrome.[202] Genetic research also identified hypogonadotropic hypogonadism and certain types of mental retardation as a gene defect.[203,204]

Clinical observation of families that have a history of fainting, drowning, sudden death, or other loss of consciousness events determined a heart condition known as Long-QT syndrome. An abnormal heartbeat in people with this affliction can make them lose consciousness. Therefore, if in the water, otherwise healthy children may drown. Families may be evaluated with an EKG. Those family members with Long-QT syndrome can simply take medications to avoid possible sequelae that might result in death.[205]

Scientists were able to narrow down the genetic polymorphisms that provoke spontaneous ischemic stroke in children.[206] Studies of families with frequent childhood immunodeficiency disorders within them led to identifying the genetic basis of the disease.[207]

Though identification of responsible genes has not yet led to cures for these tragic hereditary conditions, it lets expectant parents prepare for the potentiality of illness in their offspring, and provides science with avenues for better understanding the diseases' structure. Gene transfer may someday allow us to cure these diseases. For instance, surgeons have had some success taking thymus tissue discarded from infants undergoing heart surgery, culturing it and transplanting it into complete DiGeorge syndrome patients.[208] (Complete DiGeorge syndrome occurs in infants with "profound" T-cell deficiencies due to an incomplete thymus.) We have no more worthy objective than uncovering the causes and cures of childhood disease, via any *logical* means. Our argument is that the animal model, with parity between species so startlingly low, is not logical, and certainly not scientific.

Part of the Problem

The "system" that perpetuates and sustains the use of animal models in pediatrics, as elsewhere, is unrelentingly vigorous. The Yerkes Regional Primate Research Center is a facility where they farm, sell, and conduct experiments on a variety of primates. Researchers there busily document similarities between humans and these primates to justify their revenue supply. For instance, they point to macaques and sooty mangabeys that abuse their children as a model for human child abuse.[209] To what end are these studies? Surely there are sufficient abusive human families from whom to garner statistics.

Unfortunately, neither our government nor charities is in accordance. The FDA perpetuates the charade that animal tests prevent human birth defects and lead to cures for childhood disease. The March of Dimes is one of the largest charitable funders of research on childhood disease. It originally raised money to prevent polio. After the polio vaccine was developed, March of Dimes modified their course and began distributing grants for purposes of eliminating birth defects. Tracking the monies the March of Dimes allocated, and the outcome of the research, reveals ambiguous results.

On the one hand, we can look with gratitude to the March of Dimes for numerous achievements. Clinical researchers supported by March of Dimes reported the association between the acne drug Accutane and birth defects. This led the FDA to begin a clinical study that confirmed the reports. The use of ultrasound to detect heart abnormalities was also, in part, funded by March of Dimes. (Interestingly, the use of ultrasound to diagnose heart abnormalities is now used in animals, but it was developed in humans.) Sickle cell anemia research led to another great discovery—the use of amniotic fluid sampling to diagnose diseases in unborn babies while *in utero*. March of Dimes supported Dr. Yuet Wai Kan and colleagues, developers of amniocentesis.

However admirable the March of Dimes's mission and the developments that have issued from its funding of nonanimal studies, the March of Dimes is notorious for spurious allocations, some of which we have already characterized. A few more of the innumerable examples follow: March of Dimes funded an experiment observing gonad development in opossums decapitated after birth.[210] Researchers compared this with other studies done on gonadal development in marsupials. The organization funded research on isolated sheep uteruses and rabbit hearts.[211,212] Also underwritten was research where lambs bled to death.[213] Results of these bleeding experiments conflicted with those garnered by bleeding to death rabbits, cats, dogs and rats, again highlighting the differences between species. Researchers have looked at the effects of Valium-like drugs on nonhuman fetuses.[214,215] Whilhite and colleagues, funded by March of Dimes, studied the effects of retinoids on hamster, rat, mice and dog fetuses. They reported that the effect they were studying varied from species to species.[216,217]

A particularly meretricious aspect of the argument in favor of animal experimentation concerns Nobel Prize winners funded by the March of Dimes. Lobbyists would have us believe that the March of Dimes funded scientists who won the Nobel Prize for having experimented on animals. However, this simply does not follow. Everyone in the medical community, including us, has done animal experimentation, as we have explained previously. Only recently could one get into and through medical school without performing animal experiments and some medical schools still demand it. So it is true that most Nobel laureates used animals at one time or another. Nonetheless, this does not mean that animal experimentation was necessary or that Nobel-winning achievements depended on it. The Nobel Prize is awarded for specific contributions. If a person who had never published a paper cured cancer tomorrow, he or she would win the Nobel, irrespective of previous experience. Conversely, many scientists who have conducted hundreds of experiments and published hundreds of papers never make a significant contribution to science. March of Dimes propaganda notwithstanding, receipt of the Nobel Prize does not rely on animal experimentation.

The Future of Studying Childhood Diseases

Dr. Neal Barnard summed up the futility of animal experimentation for children: "The animal researchers would have us look at a drowning child in the water. We are with a hundred dogs in a boat. They suggest we throw the dogs into the water so they can die with the child."

Understanding the data presented in this chapter, one can appreciate the truth of Dr. Barnard's assessment. And whereas the animal model is irrelevant to childhood diseases, many other modalities are not. We are far from without options in our quest to protect our children. Human clinical studies, clinical observation, birth defect registries, epidemiological studies, autopsies, test tube research and other ways exemplified in this chapter provide tangible, usable data that will benefit human babies.

While it may be difficult for parents who have lost children to agree to autopsies, many concede because it is evident that this sort of research will help prevent subsequent tragedies. Among the diseases illuminated by autopsy are SIDS, heart disease, fetal alcohol syndrome, retinopathy of prematurity, anemia, cancer, cystic fibrosis, infantile kernicterus, erythroblastosis fetalis, neonatal giant cell hepatitis and biliary atresia, hyaline membrane disease (infantile respiratory distress syndrome), and many more.[218] Autopsies found evidence that maternal hypercholesterolemia may cause atherosclerosis in the child.[219] Post-mortem examinations are invaluable. In combination with epidemiology, astute clinical research and *in vitro* research, autopsies raise the overall prospects for children's health.

Sensible but controversial, examination of naturally aborted fetuses is also an obvious source of good data. Babies who do not make it to term can reveal a lot about hazardous toxins and genetic defects to which other developing fetuses may be susceptible. Researchers have also studied fetal development in women who are about to have hysterectomies. As mentioned previously, substances pass into the placenta at different rates in different species. Therefore, human placentas are useful for toxicology research. By using discarded human placentas, scientists garnered valuable information about the transfer of medications across the placenta.

Until recently, amniocentesis was necessary for determining Rh-factor compatibility. Incompatibility requires treatment to avert hemolytic disease. Now, *in vitro* research is refined enough that scientists can derive the same conclusions from blood samples. A more comprehensive test, called "five-color fluorescent in-situ hybridization analysis," will soon provide a non-invasive method of testing for defects.[220] By funding the development of this more advanced biotechnology, the government and charities move science forward.

As in other fields of medicine, prevention is too often overlooked as a means of lowering the incidence of disease and death. A recent study suggests that actual medical conditions account for only twelve percent of childhood deaths. That means that most deaths in childhood—eighty-eight percent—are preventable. Such deaths account for the bulk of mortality in children between the ages of one and nineteen. To decrease this frequency, the Public Health Policy Advisory Board recommended the following:

- Emphasize the central role of family and community
- Address disparities in healthcare
- Recognize homicide and suicide as public health emergencies
- Reduce alcohol and drug abuse
- Reduce availability of firearms
- Establish child death reporting systems for unintentional deaths
- Address the increase in incidence and mortality of childhood asthma

Research, even good research that supplies applicable data, is only one aspect to decreasing birth defects and other childhood diseases. Benjamin Franklin had a point when he stated that an ounce of prevention is worth a pound of cure. As emphasized throughout this chapter, much of what epidemiology brings to light points to preventable exposures, conditions that introduce toxic substances or organisms to the embryo or to small children during critical growth phases. Animal experiments, which merely reiterate human data or are unreliable, do not hasten an end to these serious diseases. The end is in sight and within reach—*prevent exposure*. However, the glamour in biomedical research is in searching for a "cure," and it is that carrot that keeps those in lab coats on the chase. The work of prevention is no fun, not intellectual enough for some, and will never get headlines. But prevention can keep children from becoming painfully ill, before and after birth, and it can prevent eighty-eight percent or more young people from dying.

Real progress could be made *now* to avert smoking, drinking alcohol, drug use, teenage pregnancy, and insufficient prenatal care if monies presently allocated to animal models were redirected to education and outreach. Prevention would not only mitigate the suffering of innocent babies, but would also take a great burden off the medical-care system. The financial savings to society and emotional well-being would be profound.

chapter 7

Diseases of the Brain

Notably in the identification of central nervous system activity, animal models are unreliable indicators . . . some drugs of proven value in man have negligible or paradoxical activity in laboratory animals . . . inconsistencies are an inevitable outcome of fundamental species-determined differences: and doubtless a number of compounds of potential therapeutic value are lost to medicine, having demonstrated little activity in an array of inappropriate animal models.

—Dr. J. F. Dunne, *Textbook of Adverse Drug Reactions*

Nothing is more devastating than a diagnosis of a progressive neurologic disease, since, whatever the condition is, it will worsen. The few treatments available usually treat the symptoms but not the disease. The tragedy of neurologic disease is compounded by the fact that it attacks the very characteristics that make us who we are: our ability to think, remember, and move, and our personality. The nervous system's complexity and our still rudimentary knowledge of its function challenge scientists daily. As a result, we have few means to deal with neurologic deterioration as more and more people are affected by it. That number will continue to increase because science has arrested so many other diseases, to which people would have succumbed in previous years.

Disorders of the nervous system have many origins—birth defects, poisoning, infection, metabolic defects, vascular disorders, inflammations, tumors, degeneration and injury. As is true in other medical disciplines, researchers continue to induce the symptoms of these disorders in animals, and then scrutinize animal brains and nervous systems in an attempt to fathom treatments and cures. Unfortunately, animal studies have done little to elucidate the underlying neurological mechanisms in humans. Human research however, has.

Logically, the practice of carefully scrutinizing the human who died from the actual disease is straightforward, informative, and reliable. Autopsy makes sense. Early autopsies characterized the anatomy of the brain and how that anatomy related to function. Gradually, scientists came to understand that the scope of conditions that could affect the brain was very broad. They could distinguish the cerebral abscesses, aneurysms, tumors, the changes of multiple sclerosis, congenital defects, and other lesions. Autopsies proved that brain damage on one side led to paralysis on the opposite side of the body.

Microscopes of ever-greater magnification enhanced the value of autopsy. Researchers were able to catalog symptoms and explain them in light of the pathologic changes seen in the nervous system. Scientists can now identify the causes of afflictions of an infinitesimal size such as the newly discovered prions that bring on variant Creutzfeldt-Jakob disease. German neuropathologist Alois Alzheimer, who first described the disease that bears his name, contributed significantly to the process of neurology because he standardized the way autopsies on the nervous system are conducted and developed dyes for staining nervous-system tissue.

Also of tremendous significance were autopsies of individuals with supposed neurologic diseases in which no pathologic abnormalities could be identified. Personality disorders were characterized by the lack of pathologic changes found at autopsy.[1] Prior to that, they had been thought to be due to pathological changes in the brain. Despite the predilection for animal models, neurology still depends heavily on the autopsy. Note what pathologists R. B. Hill and R. E. Anderson say about this:

> In recent years, participants in meetings of the American Association of Neuropathologists have heard criticism about the increasing use of animal models to study human neurologic disease.... A strong cadre of diagnostic and research neuropathologists believe that only human material can provide relevant answers to many problems about human central nervous system disease. In fact, examination of the data bears out this contention. Of the 185 abstracts presented at the 1985 meeting of the American Association of Neuropathologists, 115 (62%) were presentations of human neuropathology, and an astounding 81 (43%) were based on investigations of human brains at autopsy. Among these autopsy studies were seven presentations of either the first complete description of a newly recognized human disorder, or one of the first complete descriptions of an uncommon human neurologic disease.[2]

Today, *in vitro* analysis of autopsied human tissue is supplying even more detail. And with modern imaging technology, we need no longer even wait for human autopsy to find out what is wrong and where.

Among the technologies used to map brain function and to identify brain damage while the patient is still alive are: Magnetoencephalography (MEG), magnetic resonance imaging (MRI), functional MRI (fMRI), magnetic resonance spectroscopy (MRS), positron emission topography (PET), single-photon emission computed tomography (SPECT), event-related optical signals (EROS), and transcranial magnetic stimulation (TMS).[3] These many other modalities aid us in addition to our clinical observations and autopsies.

How the Brain Works

What we know so far is that the brain's most important function is receiving, organizing, and transmitting nerve impulses. Like vehicles along freeways, nerve impulses from throughout our bodies deliver information to and from our brains. They move through two intersecting networks, the *central nervous system* and the *peripheral nervous system*. The central nervous system is "central"—the brain and spinal cord being the core. The peripheral nervous system fans out from the spinal cord in a branching array. Bunches of *neurons* (nerve cells), tightly wrapped in connective tissue like insulated electrical wires, spread to the organs and throughout the rest of the body, down to fingertips and toes.

A neuron is a long cell. On one end are bushy dendrites that convey messages toward the nucleus, and on the other end is a long axon that conducts messages away from the nucleus. The passing of messages from one neuron's axon to the next neuron's dendrites is the transmission process. Messages are carried from neuron to neuron by neurotransmitters. When functioning optimally, each neuron generates and emits a new batch of neurotransmitters when it receives the signal from a nearby neuron. Each neurotransmitter emission ordinarily binds to sites on the outside surface membrane of adjacent neurons. The sites are called *receptors*. This binding precipitates an electrical change within the neurons, and as the change moves along the length of neuron, it eventually produces more neurotransmitters to impart to the next neuron. It is like a relay race, each neuron picking up neurotransmitter messages and passing them to the next neuron through new neurotransmitters.

The relay transmission between neurons requires both electrical and chemical events. Neurons inherently seek a polarized state. What this means is that the neuron has a negative charge inside the cell, because positively charged potassium ions move freely through the cell membrane and larger negatively charged molecules cannot. Positively charged sodium ions are kept entirely outside the cell until a stimulus occurs in the form of neurotransmitters occupying receptors on the cell surface in the dendrite area. The branching of the dendrites increases the cell's surface area. Once received at the dendrites, the signal is then passed on toward the cell nucleus. As the signal moves, the cell membrane becomes permeable to sodium. Once sodium enters the cell, the cell's charge becomes positive, and this detonates an exchange of ions all the way down the axon. Thus begins a depolarizing process, reducing the negative charge. The ion exchange gradually dissipates as the high internal sodium concentration pumps first potassium and then sodium ions outside the cell.

Just like an exhausted relay runner, the neuron reverts to its polarized state, but not before passing the message to the next neuron. This happens when the electrical signal reaches the tip of the axon. The signal stimulates small pouches, called *vesicles,* to release neurotransmitters out into the submicroscopic space between neurons (synaptic cleft). Here, the new batch of neurotransmitters attaches to receptors on the dendrites of another neuron, which propagates a depolarizing action of its own.

Understanding this passing of messages from one neuron to another is a vital part of neurology. Another aspect is identifying which part of the brain is responsible for which function. Over the eons, observation and autopsy examination of patients with brain damage created a reservoir of information about which parts of the brain are responsible for which activity. Even in the early fervor of animal models precipitated by Claude Bernard, *The Lancet* pointed out in 1883, "It is an interesting and noteworthy fact that pathological observation is doing more to advance our knowledge of cerebral localization than physiological [animal] experiment."[4]

In 1958, Dr. Hugh Jarvie reviewed the way in which science had since garnered knowledge concerning the location of various brain functions. He concluded:

> I cannot help feeling that the answers to our questions about the functions of the human brain will not be found in that way [animal experiments], but as they always have been found—in the careful collection of clinical facts and their pathological correlations.[5]

Injuries

Documentation exists on many hundreds of thousands of human brain injuries that affected mobility, function, and behavior. Until PET, MRI, and CAT scans, these copious data provided our most accurate picture of brain-activity regions. Brain injury stories can be pretty bizarre. For instance, patients who injure specific parts of the cortex called the *parietal* and *occipital* lobes suffer from a disorder known as *prosopagnosia.* Prosopagnosia patients cannot discern one face from another, but can recognize old friends by their voices. In another example, patients injured in the area of the cingulate sulcus lose the ability to care for themselves. Temporal lobe epilepsy, usually acquired from brain injury, produces hallucinations of a religious nature, which can be interpreted as visions.

Brain damage was the suspected source of speech disorders. In 1807, a scientist named Gall theorized the location of some of the speech centers in the brain based on human autopsies and clinical examinations. Then Paul Broca, in 1861, published his findings based on clinical observation and autopsy of two patients, establishing the speech center of the brain.[6] Although other animals do communicate, it is safe to say that the articulation with which humans form words and sounds is unique in its mechanics and complexity. It would be ludicrous to presume that knowledge of significant value, in regard to speech disorders, could be drawn from other creatures. Jean Martin Charcot, a well-known French neurologist, commented:

> The only really decisive data touching the cerebral pathology of man are, in my opinion, those developed according to the anatomico-clinical method. . . . To it, I may justly say, we owe whatever definite knowledge we have of brain pathology. As for the localization of certain cerebral functions, this method is not only the best, but the only one that can be employed. What light, for instance, could experimentation [on animals] have thrown on the question as to the seat of the function of speech? . . . The study of the brain, if it is to bear fruit, must be made on man, i.e., at the bedside and in the post-mortem theatre; the discovery of the exact seat of aphasia was made in that way and could not have been made in any other. . . . The utmost that can be learned from experiments on the brains of animals is the topography of the animal's brain, and it must still remain for the science of human anatomy and clinical investigation to enlighten us in regard . . . of our own species; and in fact, it is from the department of clinical and post-mortem study that so far all our best data for brain localizations have been secured.[7]

Another instance of human study yielding results is that of phantom-limb pain in amputees. Phantom-limb pain occurs when a patient loses, for instance, a foot, but still feels pain in the same foot. The origin of the pain was disputed for years. Did it come from reactivation of sensory pathways that were dormant or did the brain lay down new pathways in response to the injury? Based on the study of patients shortly after amputation, it appears that the brain just reactivates old pathways causing the patient to "feel" his or her amputated limb.[8] This research would be impossible to perform on animals because animals cannot identify the source of their discomfort to researchers.

Human-based research has yielded results serendipitously as well. J. A. Vilensky and S. Gilman of the Department of Anatomy, Indiana University School of Medicine, Fort Wayne, Indiana wrote in the journal *Motor Control:*

> From the late 1800s until approximately the middle of the 20th century, neurosurgeons made discrete motor cortex lesions in humans in attempts to reduce or eliminate a variety of involuntary movements, resulting mainly from epilepsy. In some cases, the neurosurgeons tested and recorded their patients' ability to perform various movements and to perceive various types of sensory stimuli after the operation. Although these studies have been largely forgotten, they have an immense advantage over primate lesion studies for understanding the function of the motor cortex because the patients were able to attempt to perform complex movements upon request, and to describe their perceptions of cutaneous stimuli, including integrated sensations (e.g., recognition of objects by palpation alone).[9]

No doubt some of these studies were repeated in nonhuman primates and heralded as a breakthrough.

Diagnostics and treatments for all brain injuries that were modeled with animals have been very disappointing. Of twenty-two drugs developed on animals as therapeutic in spinal cord injuries, imposed of course, none worked in humans.[10] Recently, animal models once again misled physicians.[11] Human clinical observation led to the finding that a decrease in body temperature protected against brain injury if hypothermia was induced prior to the brain injury. Despite the known historical record of animal models failing to predict human response to hypothermia, researchers pursued the study of hypothermia in animal models after brain injury. The technique failed on dogs, but is used daily on humans undergoing surgical correction for congenital heart defects. (During surgery for severe heart defects, there is sometimes a risk of brain damage.) Based on success with the animal models of stroke,[12] physicians lowered the body temperature of humans brought to the emergency room suffering from brain injury. Unlike the animal models that revealed hypothermia would also have a protective affect for the injured brain, humans showed no change or actually got worse.

Although time and money have been wasted and human lives endangered because of animal models, clinical research has revealed ways to reduce the aftereffects from brain injury. Human experience has supplied life-saving measures. Brain injured patients do better if they are transported to emergency rooms as soon as possible after injury, then rapidly treated for the low blood pressure associated with trauma and hypoxia (low oxygen levels in the brain). We know that it is necessary to relieve the high pressure in the brain with surgical intervention as soon as possible.

Multiple Sclerosis

Nerve axons can be myelinated or unmyelinated. A myelinated-nerve axon has its own protective sheath, a white fatty substance called *myelin* that insulates the axon and optimizes the transmission of impulses. But in people with multiple sclerosis (MS), information deliveries gradually become less reliable. It is as though chunks of the nerve freeway disappear; some messages reach their destination, others do not. This happens because the myelin breaks down in a process called demyelination. In demyelination the myelin swells and becomes inflamed, then detaches from the fibers. When detached myelin disappears, patches of hardened or sclerotic scar tissue form over the fibers, incapacitating their signaling capability.

Multiple sclerosis (MS) tends to occur in people in the prime of life, between the ages of twenty and forty. It is twice as common in women and largely a disease of people of northern European descent. MS is the most widespread neurological disease affecting young adults.[13] The tragic condition varies in severity and progression, beginning unpredictably with vision problems and occasional tremors. Ensuing fatigue and muscle weakness, which could suggest other problems, make it difficult to diagnose. MRIs can locate the presence of sclerotic lesions on the brain and spinal cord though. Eventually, MS victims lose all control, become wheelchair-bound and often die from the disease.

MS is actually a spectrum of diseases ranging from relatively benign to totally devastating. All forms are autoimmune, meaning that the body's immune system attacks itself. This we know from studies of MS patients who had lymphocytes and macrophages in the lesions of the myelin sheaths. It appears that immune cells mistake myelin as a dangerous foreign substance and attack it. Animal models suggested that the damage stopped with the myelin, but in humans the underlying nerve tissue, the axons, is also compromised.[14] This knowledge affects how scientists approach not only the study of MS but also the treatment.[15] Having developed two immunotherapy treatments for multiple sclerosis after years of using animal models, researchers at Stanford and the National Institutes of Health were forced to curtail trials because human patients symptoms worsened or they developed allergic reactions.[16,17]

Most studies of humans have been done via clinical research and *in vitro* research using T-cell lines and cells cloned from diseased individuals.[18] Using state-of-the-art science, an international research team identified four different patterns of multiple sclerosis. Pattern one and two were both similar to autoimmune encephalomyelitis, one T-cell mediated and one antibody mediated. The last two patterns were not so reminiscent of autoimmune encephalomyelitis as of oligodendrocyte dystrophy, since damage occurred from a virus or toxin. The team concluded:

The mechanisms and targets of demyelination in MS may be fundamentally different in distinct subgroups or stages of the disease. Thus, a given therapy that is beneficial to one group of patients or at one stage of the disease may or may not be beneficial to another group.[19]

One can easily appraise the significant challenge to those who have devoted their careers to studying and attempting to stem this disease.

Millions of people worldwide have MS and over half a million are Americans, so naturally, this nightmarish problem attracts a lot of research dollars. Still, though scientists have described types of MS, the exact cause of the disease has not been elucidated, nor has a cure been found. As usual, large percentages of the available resources have been directed to animal models. Many years and many millions of dollars in the animal lab have not helped MS victims. The animal "model" of MS is called *experimental autoimmune encephalomyelitis* (EAE). First induced in monkeys in 1933, EAE has since been induced in guinea pigs, marmosets, rats, mice, and rabbits.[20] Whereas the animals do have some of the same symptoms, the cause of the symptoms is different and, just as important, imposed. Remember how dissimilar versions of MS are, even within humans. Further, there have been real problems getting EAE pathology to progress even so far as demyelination. As scientists cannot induce sickness in the animals in the same way, naturally, they haven't even begun to cure multiple sclerosis itself. Note what scientists say about the model:

In EAE, however, the inducing antigen is known, whereas the antigen specificity of the immune reaction in MS has not been identified. Furthermore, many models of EAE are characterized by perivascular inflammation without significant CNS demyelination, in contradistinction with MS in which demyelination is the primary feature. Furthermore, EAE is not a naturally occurring autoimmune disease.[21]

Those are major differences. Drs. Claude Genain and Stephen Hauser of the Department of Neurology at the University of California San Francisco, who are actually proponents of animal models, wrote,

Phylogenetic differences may limit the usefulness of EAE models, and indeed no single form of rodent EAE recapitulates all the clinical and pathological features of MS. . . . Many therapies have been found to be effective against EAE, yet therapeutic success against EAE has not been a *reliable* prediction of success against MS. As is emphasized below, other experimental results derived from rodents with EAE simply cannot be applied to the human condition. Important differences between EAE and MS also exist in terms of pathology. In most forms of acute EAE, inflammation predominates over demyelination whereas acute demyelination is the pathological hallmark of the early MS lesion [emphasis added].[22]

There are numerous other animal models of MS: the shaking canine pup, the shiverer mouse, the myelin-deficient rat. None replicates the human disease. All kinds of viruses can lead to demyelination in animals—mouse hepatitis virus, Sindbis virus, herpes virus, and Theiler's murine encephalomyelitis virus.[23] For years, Visna, a disease of sheep, has been a pet focus of the animal model community presumably at work on MS. But Visna's correlation to MS is at best tenuous. A virus causes Visna. No such organism can be isolated in MS patients. Likewise, mice inoculated with corona virus get MS-like symptoms. However, no one knows if the mouse disease would develop without direct brain inoculation. Also, it does not relapse as it does in humans. Dr. Gibbs, writing in *Scientific American* in 1993, had this to say about animal models of MS:

> To the 2.6 million people around the world afflicted with multiple sclerosis, medicine has offered more frustration than comfort. Time after time, researchers have discovered new ways to cure laboratory rats of experimental induced encephalomyelitis, the murine model of MS, only to face obstacles in bringing the treatment to humans.[24]

Why administer chemicals and viruses to animals that happen to result in loss of myelin? They do not get MS. It is like cutting out an animal's heart to simulate heart failure. These studies are supposed to show MS's causes. The end results may be the same—heart failure, loss of nerve conduction, and so forth. But since the mechanisms are not the same, the efforts are unwarranted.

While MS research on animals continues to baffle, human clinical and epidemiological studies have linked MS to environmental factors.[25] Epidemiologists have also confirmed that MS has a hereditary component. Not just one gene introduces susceptibility, but multiple genes contribute.[26] For instance, noting the prevalence of MS among people of Scottish descent, epidemiologists and *in vitro* researchers located the HLA allele *DR2*. This gene is twice as common in Scotland as in England, and definitely associated with MS.[27] One of the discoveries was that the major histocompatibility complex (MHC), a group of genes that code for distinct markers on our cells' surface that differentiate self from nonself, are involved. That in part accounts for the autoimmune aspect. Additionally, all MS patients appear to have a specific variant in the *APOE* (apolipoprotein E) gene, which codes for a protein involved in the transport of lipids. As myelin is fatty in nature, scientists speculate that the variant may be affecting the myelin negatively.[28] A recent finding suggested that herpes virus 6 might provoke MS in genetically susceptible people. Scientists studied patients with MS and found that herpes virus 6 was in the plaques that form in the nervous system of MS patients.[29,30] This may effect how patients are treated. In further efforts to characterize human MS patients, researchers have found that many have, in addition to myelopathy, atypical anticardiolipin antibodies, optic neuropathy and cerebellar syndrome.[31]

On the therapeutic side, pharmaceuticals developed using animal models never cease to disappoint. Copaxone (Glutiramer acetate) was effective in EAE animals but has shown mixed success in humans. Also there are many side effects.[32] Tumor necrosis factor was given to EAE animals and worked well, but it did just the opposite in humans.[33] Scientists found that one treatment that showed promise in animal models—intravenously injected immunoglobulins—does not remyelinate multiple sclerosis lesions any more than placebos in humans.[34] Two other MS drugs—altered peptide ligand formulas known as CGP77116 and NBI 5788 that constituted an immunotherapeutic approach— worked well in animals. However, clinical trials came to an abrupt halt after several people almost died.[35] Speaking of animal models of autoimmune diseases in a reputable immunology journal, Veena Taneja and Chella S. David stated, "Of course, it is not possible to reproduce a complete human disease in an animal."[36]

Although there is no cure for MS, different versions respond to specific treatments, mitigating the duration and severity of relapses. For years, victims took glucocorticoids with marginal success. Presently, oral or intravenous corticosteroids reduce nerve tissue inflammation and shorten the flare-ups. However, as the medications aggravate osteoporosis and hypertension, MS patients must take them judiciously. Tizanidine hydrochloride (Zanaflex) is a new oral treatment for muscle spasticity.[37] Multiple autoimmune medications have been discovered recently and tried in diseases such as MS, rheumatoid arthritis, systemic lupus erythematosus, myasthenia gravis and others. Among these, cyclosporin A, tacrolimus, and Rapamycin (sirolimus) are all from microbial origin and structurally similar. Their effect was predicted by their structure, not by animal models.

Another experimental branch of treatment are the interferons. These are genetically engineered copies of small proteins known as cytokines that occur naturally in the body. They "interfere" with viral replication and regulate the immune system. There are three kinds of interferon (IFN): α,β, and γ. Although animal studies with IFNγ looked positive, humans actually got worse on the therapy.[38] On the other hand, beta-interferon (INFβ has been used with some success in MS patients. It is made using human recombinant INF and *E. coli*. However, beta interferons do not altogether reverse neurologic disability, and there is no conclusive evidence that they prevent permanent disability. Also, they are expensive. Sometimes patients develop antibodies to them, rendering the drugs ineffective.

Researchers are pursuing other approaches to MS therapy, for example, plasma exchange. This method would remove immune factors in the blood and exchange them for new, healthy plasma. Though researchers are still not sure how much of a role herpes virus 6 has, acyclovir might be an option. Two other possibilities are immunizing against the particular white blood cells that attack the myelin and stem cell therapy.

Motor Neuron Diseases

Motor neuron diseases (MNDs) are diseases of the central nervous system. They are characterized by muscle weakness, atrophy of the muscles, and eventual paralysis due to denervation. They include amyotrophic lateral sclerosis (ALS, also known as Lou Gehrig's disease), pseudobulbar palsy, bulbospinal neuronopathy (Kennedy's syndrome), spinal muscular atrophy, and neuroaxonal dystrophy. These diseases have no cure and most of their causes remain a mystery. Efforts to emulate these little understood illnesses in the animal lab are curious, even in concept . . . and distinctly fruitless. More promising is the epidemiology that led to an understanding that some MNDs are hereditary, and genetic research that isolated the mode of inheritance and the gene carrying the disorder.

Amyotrophic lateral sclerosis (ALS), first described in 1869 by Jean-Martin Charcot based on autopsy findings, gained notoriety when baseball great Lou Gehrig was diagnosed with it. Now frequently referred to as Lou Gehrig's disease, at least one form of the disease is attributed to the gene *EAAT2* on chromosome 9, which may produce a defective protein.[39,40] This was discovered by performing autopsies on patients who had died of ALS. Researchers have used transgenic mice in an attempt to replicate findings in humans, but the results have varied depending on species and strain of animal used. Sherril Green and Ravi Tolwani, both of whom favor the animal model stated:

> Aside from non-human primates, a species in which inherited or naturally acquired MND has not been reported, none of the aforementioned animal species has a well-developed corticospinal tract, a system that is uniquely and specifically involved in ALS.[41]

Alzheimer's Disease and Dementia

Dementia is gradual loss of memory that eventually erodes the ability to conduct everyday activity. Alzheimer's disease (AD) is dementia's most prevalent form. First described in 1906 by Dr. Alois Alzheimer, AD affects about five million U.S. citizens. Its estimated cost is in excess of $100 billion annually.[42] Although aging does not necessarily cause this central nervous system affliction, Alzheimer's symptoms can increase with age. One percent of all Americans over the age of sixty-five has Alzheimer's. The disease becomes more prevalent the older the population examined, affecting between twenty and thirty-five percent by age eighty-five. Responsibility for America's roughly fifteen million Alzheimer's patients already greatly weighs on family members and care facilities.[43] Combine this with the preponderance of the disease that is boded by the ballooning elder population. Sheer numbers and apprehension have intensified research into its causes, treatments, and plausible cures.

Because many neurological diseases mimic Alzheimer's symptoms, it has always been difficult to diagnose. Until recently, the only way to determine the disease's presence definitively was at autopsy. Scientists gathered much initial understanding of Alzheimer's in this way, by looking directly at patients' brains. They found that neurons in the brain—primarily in the hippocampus and neocortex regions, had deteriorated. There were little lint-like wads called *neurofibrillary tangles* within the cells, hardened protein deposits called *neuritic plaques* outside the cells, and general pockets of degeneration called *granulovacuolar degeneration bodies*. It was evident that proteins running amok had something to do with the aberrations.[44]

Research efforts first plumbed to determine the nature of the neurofibrillary tangles, brought about by a protein called *tau*. Scientists located the tau gene on chromosome 17. Tau is important because it contributes to the formation of the all-important intercellular structures called *microtubules*. Structurally, if neurons were buildings, microtubules would be their interior walls. Microtubules impart support and encasements for conducting nutrients and cellular components throughout the cell. When tangles take the place of normal channels, it severely limits the cell's efficiency.

Concurrently, the lab-animal researchers launched into an unproductive investigation of a made-to-order transgenic mouse, a mouse that had a mutated tau protein gene inserted. However, the mouse's tau did not result in any Alzheimer's-like symptoms or even a neurological change. This implied that tau was unimportant.[45,46,47] Years passed and tau research was largely remanded to the back burner. Scientists, now knowing better, admit that the response to tau is "highly species-specific."[48] There is no successful animal model of Alzheimer's disease.[49]

Surprisingly, several clinical studies and autopsies showed that the brain changes commonly associated with Alzheimer's—the *neurofibrillary tangles* and *neuritic plaques*—do not always result in dementia. The first big breakthrough in AD came in Liverpool in 1976. Autopsying the brains of AD patients, Dr. David Bowen found depleted supplies of the neurotransmitter acetylcholine. Acetylcholine would normally help neurons communicate.[50,51] Bowen's finding and subsequent clinical observation suggested what is known as the "cholinergic hypothesis" to explain AD. This postulates the idea that neurons secreting acetylcholine (Ach), or being stimulated by Ach, are damaged, and AD ensues.[52]

That nondemented humans suffered from cognitive deficits after taking anticholinergic medications such as scopolamine reinforced this hypothesis. These results were then duplicated in nonhuman primates.[53] Again, the animal model merely reproduced finding in humans. Further, autopsies and clinical observation showed that the brains of patients with AD demonstrated less choline acetyl transferase (ChAT) activity.[54] Again, these findings were duplicated in animals. R. Bartus stated in 1986 about animal models for AD, the ". . . value of any model or approach will depend not on the inherent logic of the principle that guided its development, but on its ability to make meaningful predictions about the clinical condition it was designed to study."[55] Studies that merely duplicate known human conditions are not predictive.

The relative deficiency of Ach in the brains of AD patients stirred scientists to locate a medication to increase the amount. They knew that an enzyme named cholinesterase breaks down Ach. If the enzyme could be inhibited, slowed down, then more Ach could be available. Cognex (or tacrine), a cholinesterase inhibitor, was already in use when approved for AD treatment. Its effect was predicted based on its known properties. Pharmacologists designed a follow-on drug, donepezil (Aricept), also a cholinesterase inhibitor, especially to treat AD based on its structure. Other drugs are also in the works, designed around the structure required to inhibit cholinesterase and minimize side effects.

While the animal model community continued to try and induce AD in animals, epidemiology and *in vitro* research pushed ahead, establishing the genetic basis for Alzheimer's. The disease can be either hereditary or not. When Alzheimer's is hereditary, it is "familial." When not, it is "sporadic." The greatest clues to the etiology or cause of Alzheimer's have come from linkage studies of familial Alzheimer's disease.[56] And while epidemiologists continued defining subsets of Alzheimer's victims, *in vitro* research scientists went to work analyzing the neuritic plaque in humans, Alzheimer's second characteristic. Examining the tissue under microscopes, Dr. George Glenner learned in 1984 that the amyloid-β peptide—a short protein fragment—was a pervasive component in the plaque. The beta-amyloid precursor protein (βAPP), from which amyloid-β peptide is derived, is a lengthy protein (from 695 to 770 amino acids long) most of which is within the outer cell membrane. Additionally, a short piece of βAPP juts into the cell and a longer piece juts outside the cell. A now-identified gene on chromosome 21 governs βAPP. Because genes contain instructions for protein synthesis, mutations can mean that the protein specified will turn out differently than it should.

βAPP fragments—amyloid-β peptides—that are forty amino acids long have some as yet undetermined function and are not deleterious. But amyloid-β peptides that are *forty-two* amino acids long are dangerous. They damage nerve cells by disrupting calcium regulation, which can lead to cell death, and they may also harm mitochondria. As a result, free oxygen radicals proliferate, injuring cell proteins, lipids and DNA. As the affected cells crumble, they release compounds that may attract immune cells and an inflammatory response. The role of this general mayhem in Alzheimer's is still under discussion. What researchers noticed was that genetic mutations, sometimes on chromosome 21, interfered with the cutting of the beta-amyloid precursor protein (βAPP) and that uncut or improperly cut βAPP weave across cell membranes, sort of binding them up with stitches.

Having garnered some genetic Alzheimer's determinants, scientists wanted to know just what was transforming the normally very long beta-amyloid precursor protein (βAPP) into the very short amyloid-β peptide. Interesting developments were taking place under the microscope where researchers observed human brain cells.[57] They found that enzymes called secretases (β-secretase and γ-secretase) cleave the small peptide. In the autumn of 1999 four separate companies confirmed that the genes coding for β-secretase were presenilin 1 and presenilin 2, on chromosomes 14 and 1 respectively. Each used a different method for finding the gene.[58] None relied on animal models.

As it happens, about fifty percent of people with familial Alzheimer's disease—which again is an inherited, early onset form of Alzheimer's—have a mutation in the presenilin genes (genes that code for the protein presenilin) that predisposes them to the accumulation of characteristic amyloid plaques between neurons in their brains.[59] The presenilin codes for β-secretase, which cuts βAPP. Of course, the mutation is not dominant in all carriers. Gene studies found that approximately five percent of all Alzheimer's cases can be linked to genetic mutations occurring at the genes coding for beta-amyloid precursor protein, presenilin 1, and presenilin 2 (on chromosomes 21, 14 and 1). To "validate" this finding, research with mice was immediately under way.[60] But why, as it had already been established?

Here, an interesting lack of congruity between animals and humans again bungled the stream of revelation. Man-made mutations in mice can cause them not to produce presenilin 1. And as a result, the mice produce no amyloid-β. But in humans with presenilin 1 and 2 mutations, it does the reverse.

James M. Conner and Mark H. Tuszynski of the University of California, San Diego, summarized the frustration of animal models of AD in 2000, in *Central Nervous System Diseases: Innovative Animal Models from Lab to Clinic*:

> In the case of AD, good animal models have been difficult to come by. The full spectrum of the biochemical and pathological abnormalities characterized by AD have not been found to occur spontaneously in any animal species other than the human, and the complexity of the disease has made it difficult to generate animals with a full range of experimentally induced AD pathological alterations.[61]

In humans with presenilin mutations, cutting increases and with it an overproduction of amyloid-β peptide, especially of the destructive longer version, results.[62] If this cleaving could be stopped, perhaps Alzheimer's could be arrested, they reasoned. *In vitro* research having isolated the human gene that codes for β—secretase, the enzyme responsible for one of the chemical reactions in Alzheimer's, it may now be possible to design drugs that will interfere with this enzyme and thus prevent the disease.[63] According to an article in *Nature,* the identification of the beta-secretase enzyme:

> means that the path towards specific inhibitors is now set, and it is time not only to test the amyloid hypothesis *('in vivo veritas'),* but to find a way of halting this dreadful disease, according to Drs. Strooper and Konig, of the Flanders Interuniversitary Institute for Biotechnology in Leuven, Belgium, and Bayer AG in Wuppertal, Germany, respectively. Dr. Gurney and his team at Pharmacia and Upjohn make the observation that. . . . Asp2 is a new protein target for drugs that are designed to block the production of amyloid beta-peptide.[64]

Unfortunately, one of the secretases is a vital cellular protein also found in the pancreas. This, of course complicates the pursuit of Alzheimer's medications. It is imprudent to decrease secretase too much, as it allows the pancreas to function normally.[65] A novel protein, nicastrin, has recently been established as an integral part of modulating the production of amyloid-β. Nicastrin and presenilin combined make γ-secretase. Its discovery presents another possible alternative target for Alzheimer's drugs.[66]

Researchers have isolated additional genes believed to contribute to AD. The apolipoprotein E, or *APOE-e4*, plays a part in determining when familial Alzheimer's will manifest. Approximately twenty percent of all dementia patients have the *APOE-e4* gene, located on chromosome 19, and having the gene about doubles the chances of being affected by dementia.[67,68] APOE transports cholesterol in the bloodstream and is involved in cellular repair and regeneration. *APOE-e4* may compete with the amyloid-β peptide for removal from the space between cells, leaving a plethora of the peptide to cause trouble.[69] In separate epidemiological research, scientists found another Alzheimer's-related mutation by studying afflicted Icelandic patients and their relatives.[70] Although identifying the genetic basis for young age-onset AD was fairly easy, finding the genes involved in old age-onset was more difficult. Again, clinical research gave the answer. By studying humans, scientists were able to find a gene on chromosome 10 that contributes to the later onset version of AD.[71]

Back to the neuritic tangles brought about by tau. Again, enzymes gone awry are to blame. Some enzymes associated with the tau proteins are the casein kinase-1 (CK1) family of protein kinases. Kinases catalyze phosphorylization, which may play an essential role in tangle formation. When the researchers compared Alzheimer's-damaged brain tissue with normal tissue samples (both from humans), they found that the concentration of one of the CK1 proteins—delta kinase—was thirty times higher in the diseased tissue, another was nine times higher and the third was more than twice as high.[72] Errors in the tau gene interfere with the way tau makes microtubules. As a result excess tau builds up, instead of forming neat microtubules, and the cellular transport mechanism is disabled. More research revealed that chromosomes 1, 12, 14, 19, and 21 contribute in about half of all Alzheimer's patients. The four genes on these five chromosomes are the presenilin 1 gene, the presenilin 2 gene, the beta-amyloid precursor protein (*βAPP*) gene and the *APOE-e4* gene.[73,74,75]

Human clinical studies combined with autopsy data revealed that levels of amyloid-β, the forty-two amino acid long peptide, elevate earlier in the course of disease, even before tau proteins manifest.[76] Researchers confirmed Alzheimer's abnormalities studying the chromosomes of Down's syndrome patients.[77] Without understanding the exact methodology, it seems likely that abnormalities in *βAPP* and β-amyloid peptide catalyze a series of mistakes, of which altered tau is one, and dementia is the outcome.[78] As for the sporadic or nonfamilial form of the disease, again human-based research found the genetic link. The bleomycin hydrolase gene was mutated.[79] In nonfamilial cases, a variant of the gene encoding another protein, alpha-2-macroglobulin (A2M), predicts whether the disease will develop at all.[80] Interleukin-1 gene polymorphisms also coincide with Alzheimer's disease suggesting another target for protection against and treatment of the disease.[81]

No question, the path to a cure for Alzheimer's disease is very complex and varied, and researchers have a challenging task of unraveling its mysteries either to avert it or cure it. AD presents a dizzying array of anomalies, which science is revealing in ever-greater detail. Considering the microcosmic detail, it certainly does not make sense to waste money on animal models, when much more accurate and comprehensive information is available *in vitro*, studying humans, or on computers. In the meantime, we need no longer wait until death to determine whether a patient has Alzheimer's. Biotech has spawned new diagnostic techniques. German scientists created a method for confirming the presence of Alzheimer's using fluorescence correlation spectroscopy to find the amyloid deposits in cerebral spinal fluid.[82] In autopsy of Alzheimer's victims, deficits of two neurochemicals—somatostatin and corticotropin-releasing factor—are well recognized. Efforts are under way to diagnose their deficiency as an early marker of the disorder.[83]

Another very exciting diagnostic technique is in development. The noninvasive functional MRI (fMRI) determines dysfunction in the entorhinal region of the hippocampus, the brain's key structure for controlling memory and the prime area affected with Alzheimer's. This will be especially beneficial in detecting the disease early, when symptoms are not so pronounced.[84,85,86] The lower the blood flow in the posterior regions of the brain, the more rapid the deterioration. A new noninvasive brain perfusion imaging called SPECT can predict survival rates and increase the likelihood of a correct diagnosis.[87,88]

As for cures or vaccines, there still are none. Labs continue to try out therapies on assorted animals even though many knowledgeable scientists feel as does Dr. D. Lindholm of Uppsala University in Sweden who stated, "There is no . . . good animal model for the [Alzheimer's] disease process characterized by a loss of cognitive functions and memory decline."[89] Last year Dale Schenk and his colleagues buoyed the families of Alzheimer's patients with news of a suggested anti-Alzheimer's vaccine. They worked with mice that overexpress a mutant form of the human amyloid precursor protein, amyloid-β, which in mice brings on an Alzheimer's-like condition called cerebral amyloidosis. Their prevention of this cerebral amyloidosis with vaccination was first published in *Nature* and appeared in news media throughout the world.[90] In an accompanying editorial in *Nature,* David Westaway and Peter St. George-Hyslop stated:

> All of the current animal models (based on overexpression of human APP and/or presenilin-1 transgenes bearing missense mutations associated with Alzheimer's disease) provide only a partial model of the human condition. So, although these animals accumulate increased levels of A-β (β-amyloid? in the brain) and have many amyloid plaque deposits, they have only subtle behavioral and electrophysiological deficits. More problematically, these animals do not develop neurofibrillary tangles or show significant neurodegeneration.[91]

Optimism over the Schenk group's findings is entirely premature and misleading for this and other reasons, and because translating these results to humans is going to be very difficult and possibly dangerous. The protein used in the study may result in antibodies that damage healthy human tissue. One of the proteins that the vaccine attacks is found in human platelets. Moreover, the underlying premise of the study—that amyloid is responsible for Alzheimer's—may be wrong. This is still controversial. Tau may be the most important factor. As mentioned, not all people with amyloid plaque have Alzheimer's. Autopsies indicated that many elderly have substantial amounts of amyloid in their brains without having been demented or even mentally slower than normal before their death. Whereas amyloid is present in Alzheimer's patients, it is not necessarily the *cause* of their condition.[92]

The mice used to study the vaccine were genetically modified. But even with the altered genes, the mice still failed to reproduce the human disease. Paul F. Chapman, writing in *Science,* stated:

> Although many of these features [features of human AD] have been reproduced in one or more lines of transgenic mice, no one model has captured the full range of pathology. This might be seen as an inexplicable setback. But it can also serve as an opportunity to understand the contributions of individual pathological features to cognitive loss.[93]

Chapman's views suggest that once again animal-model adherents are trying to have it both ways. They say that they must use intact animals since only then can they see how the whole organism interacts. But when the model fails, they say that it is still a triumph since they can now separate the intact organism into smaller parts and study the parts that comprise the whole. This inconsistency is the hallmark of someone whose livelihood depends on a principle that does not work.

The mouse models were heralded as great successes again at the end of 2000 in *Science* because the mice exhibited signs of decreased memory.[94] Since patients with AD also suffer from decreased memory, the mouse models were said to mimic the human condition. This is specious reasoning at its finest, again typical of the animal model community: create a model that has one or more things in common with a human disease and based on these commonalities call the animal model a success. Logically though, just because P follows Q does not mean that P caused Q.

Other human-based data inform the picture. Approximately one-third of all stroke patients are left with some degree of dementia.[95] Stroke prevention could, therefore, lead to the prevention of Alzheimer's-associated dementia. So, clinical trials are under way to see whether antioxidants such as vitamin E or nonsteroidal anti-inflammatory drugs such as ibuprofen alleviate the toxic effect of β-amyloid.[96] Clinical studies revealed that women suffering from Alzheimer's disease improve their memories by wearing an estrogen patch. Estrogen replacement therapy has been shown to reduce women's risk of suffering from the disease.[97] Clinical research found that inadequate supplies of B_{12} and folates increase the likelihood of developing Alzheimer's.[98] Studies confirmed a correlation between aluminum sulphate exposure and memory loss.[99] However, it does not necessarily bring on Alzheimer's.

Stanley Prusiner, who won the Nobel Prize in 1997 for his contribution to our understanding of encephalopathies, believes prions may be at the root of Alzheimer's, Parkinson's, and amyotrophic lateral sclerosis (ALS). "In all of these neurodegenerative diseases there are abnormal protein deposits," Dr. Prusiner pointed out in a plenary address at the annual meeting of the American Neurological Association. He explained that prion infection causes proteins in the brain to change from a normal spiral formation to an abnormal formation and suggested that prion-induced protein changes can be seen in the lesions characteristic of Alzheimer's disease.[100]

Sometimes studies of dementia reveal non-Alzheimer's forms. Mutations in the neuroserpin gene have recently been coupled with a newly recognized type of familial dementia. Researchers noted that, "Serious consequences of neuroserpin instability are relevant to the dementias as a whole because of the much greater risk that protein accumulation poses to long-lived and non-dividing cells such as neurons." The findings answer the question of ". . . whether protein deposition and accumulation is in itself sufficient to explain the late-onset dementia" such as spongiform encephalopathies and Alzheimer's disease. The researchers go on to suggest ". . . that inhibitors of protein polymerization may be effective therapies for this disorder and perhaps for the other more common neurodegenerative diseases."[101]

Researchers studied types of dementia that reoccur within families. When dementia is inherited in an autosomal-dominant fashion, they can isolate the genes. Women who had suffered strokes, in addition to the brain changes, were eleven times more likely to be demented. Epidemiological studies linked dementia to a high fat/high cholesterol diet.[102] They also led to the discovery that individuals who smoke are twice as likely to develop dementia as those who do not. In vitro, epidemiology, and clinical research continues to inform discoveries of this nature, rendering animal-modeled re-creations a complete mockery.[103]

Parkinson's Disease

Parkinson's Disease (PD) is a ruinous illness, which mostly affects middle-aged adults. As more than a million people in the United States suffer from the disease, Parkinson's ranks second only to Alzheimer's among neurodegenerative afflictions.[104] Parkinson's musculoskeletal symptoms are apparent in clinical observation. First, there is a tremor, or shaking motion, in the hand and arm characterized by a symptom known as "pill rolling," since the patient appears to be rolling a pill between his or her fingers. Another symptom is "cog-wheel rigidity." Stiffness makes the joints' motion jerky, as a wheel with cogs, catching and moving in a very mechanical manner. Body rigidity, addled movement, smaller handwriting, weakness, and imbalance are indicators. Unfortunately, the symptoms are not just confined to the musculoskeletal system. Psychiatric problems, such as depression and psychosis, may also accompany the disease. These and muscular symptoms usually worsen with time.

Autopsies of Parkinson's patients exposed degeneration in the midbrain region called the *substantia nigra*, a deeply pigmented portion in the basal ganglia. In 1960, researchers Oleh Hornykiewicz and his colleagues at the University of Vienna performed autopsies on patients who had died with PD and found the nigrostriatal pathway had degenerated and had very little dopamine. The neurotransmitter dopamine (DA), which regulates movement and emotion, is the chief carrier of nerve signals to this region, and *in vitro* research on brain tissue of Parkinson's victims indicated dopamine deficiency. Levels of the substance that dopamine is metabolized to, homovanillic acid, were also subnormal. Researchers determined that nerves containing dopamine die in Parkinson's patients, leaving the brain without enough dopamine to function normally. Hornykiewicz and his colleagues administered a dopamine precursor and saw immediate impressive results. This changed the way PD was treated as well as opened the way for new research on schizophrenia, epilepsy, and other neurological diseases. John Hardy of the Mayo Clinic stated that Hornykiewicz's research, "fundamentally changed how neuropharmacology is practiced."[105]

In 2000, the Nobel committee passed over Hornykiewicz and awarded the Nobel Prize for Medicine or Physiology to Dr. Arvid Carlsson of the University of Gothenberg in Sweden and two New Yorkers, Dr. Eric R. Kandel of Columbia University and Dr. Paul Greengard of Rockefeller University. The three won the award for their research in neuroscience, which was mainly *in vitro* research, but they did do animal experimentation and use some animal tissues. This was research that Oleh Hornykiewicz had pioneered in humans. The oversight suggested prejudicial bias on the part of the Nobel committee. Does it really believe that animals are more like humans than humans? More than 250 neuroscientists signed a letter to the Nobel committee condemning the committee for their neglect of Hornykiewicz.[106] A close examination of the Nobel Prizes awarded to researchers who used animal models reveals many such instances.

No animal, aside from humans, suffers from Parkinson's. Researchers have reproduced the symptoms in animals and this has resulted in their receiving grant money to study them. These Parkinson's models include: the reserpine model, neuroleptic-induced catalepsy model, tremor model, models with degeneration of nigrostriatal dopaminergic neurons induced by using the chemicals 6-OHDA, methamphetamine, tetrahydroisoquinolines, β-carbolines and iron, and most recently the "parkinsonian baboon." Timothy Schallert and Jennifer L. Tillerson of the University of Texas state in *Central Nervous System Diseases: Innovative Animal Models from Lab to Clinic*:

> Many types of animal models of Parkinson's disease are available. Selecting a clinically predictive one is crucial, but has always been difficult. . . . Although some researchers have argued that only primate models . . . [are useful], there is no consensus or current bias to suggest that one species is more predictive than another in the transition from research to patient.[107]

In 1983, researchers learned from clinical observation that the chemical MPTP (a by-product of synthetic heroin) could cause Parkinson's-like damage. The fact that MPTP causes PD-like symptoms was found accidentally after drug addicts injected themselves with MPTP-contaminated chemicals. The MPTP destroyed the dopamine-producing cells in the substantia nigra. So, researchers treated mice and other lab animals to infusions of MPTP. The data was merely demonstrative. Jay S. Schneider, who has worked at both Thomas Jefferson University and Drexel University, received public funds to inject MPTP into cats' brains and into two varieties of macaques. In 1997 Schneider wrote:

> Some monkeys [injected with MPTP] had cognitive deficits and no motor deficits. Other monkeys had full parkinsonism that was produced after short-term high dose MPTP exposure, and some monkeys had full parkinsonism after long-term low dose MPTP exposure.

Inconsistent, *coerced* rather than *acquired* symptoms, which in aggregate Schneider calls "parkinsonism," are hardly a true model of this still little understood human disease. Simranjit Kaur and Ian Creese of the Center for Molecular and Behavioral Neuroscience, Rutgers stated in *Central Nervous System Diseases: Innovative Animal Models from Lab to Clinic*:

> The best model of PD to date, is the 1-meth-4-phenyl-1,2,3,6-tetrahydropyridine (MPTP)-lesioned marmoset. The neurotoxicity produced by 1-methyl-4-phenylpyridinium ion (MPP^+), a metabolite of MPTP, is thought to mimic human PD. MPTP reduces the levels of dopamine and its metabolites in the striatum. However, this model is far from ideal owing to the high cost of using primates

(MPTP is ineffective in rats) and unlike human PD, which is progressive, the neurotoxic damage produced by MPTP is reversible. Other models used to mimic PD are: (1) lesions produced by the selective neurotoxin 6-hydroxydopamine (6-OHDA) in rats; (2) akinesia induced by reserpine via dopamine depletion; (3) catalepsy induced by neuroleptics by blocking dopamine receptors; (4) lesions of dopaminergic terminals produced by the systemic administration of very high doses of amphetamines. Each of these models, although serving a purpose in PD research, does not reproduce the human condition fully. For example, the 6-OHDA-lesioned rat model does not always look overtly Parkinsonian, and with unilateral lesions the only indication of cell damage is the presence of rotations after treatment with an antiparkinsonian agent. The depletions produced by reserpine are not restricted to dopamine and there is no destruction of neurons, again unlike human PD. Also, the akinesia produced by reserpine is reversible. In these models, it is difficult to ascertain the roles played by individual dopamine receptors in the pathology as well as the therapy of PD.[108]

And is the ability to give animals the Parkinson's-like agitated movement in any way helping real Parkinson's victims? These animal models can only reproduce data already gleaned from humans. Their predictive value is negligible. The problem is, that while it is relatively easy to kill the substantia nigra with these substances and make animals shake, the reason the cells in the substantia nigra die in the animal is not the same as in Parkinson's. Furthermore, none of the models reflects the *progression* of Parkinson's symptoms as it occurs in humans.

Philippe Hantraye of Service Hospitalier, France, wrote:

However, it is essential to understand that animal models only represent an imperfect replica of human disorders [of PD], and this is so for several reasons. First, animal models are generally developed in beings (rodents, non-human primates) that are subjects with behavioral repertoires and anatomical characteristics very different from humans. These species differences are known to play a role in the clinical expression as well as in the cellular specificity of the lesions. For example, striatal degeneration in humans is frequently associated with dyskinesia, whereas in rat or non-human primates, striatal excitotoxic lesions alone are not sufficient to induce dyskinesia or chorea. Second, in addition to these species differences, the time course evolution of the nerve cell degeneration, which normally evolves over several years in neurodegenerative diseases in humans is for practical reasons, being replaced over a much shorter period of time in animal models.[109]

Echoing these views is James A. Temlett, of the neurology unit of the University of the Witwatersrand Medical School in South Africa, who stated:

Acute parkinsonism models [animal models] have limitations when compared with chronic disease states, and caution should be present when comparing parkinsonism data with human disease. . . . Animal models, including the MPTP-lesioned nonhuman primate data, but especially rodent models, are 'acute parkinsonian models' that provide more controlled conditions, preservation of tissue biochemical or molecular estimates. However, these do not reflect the complexities of the human basal ganglion.[110]

And Erwan Bezard et al. wrote in *Review in the Neurosciences*:

Discrepancies have been reported several times between results obtained in classic animal models and those described in PD [human Parkinson's], and it would seem probable that such contradistinctions can be ascribed to the fact that animal models do not, as yet, reproduce the continuous evolution of the human disease.[111]

Physicians serendipitously discovered that the nightshade plant partially relieves some Parkinson's symptoms. The only therapy for many years, it decreases acetylcholine, the chemical that is in relative excess due to the decrease of the chemical dopamine. The truth is that we will not make real progress in treating Parkinson's until we know *why* the substantia nigra dies. Once the cause is arrested, the symptoms will stop.[112] The obvious interim solution is to replace the dopamine; however, a built-in condition called the blood-brain barrier stands in the way. This protective layer inhibits many substances, such as toxins, from entering the brain from the bloodstream. In this case, it also prevents medications from reaching their destination. Dopamine cannot cross the barrier.

Through clinical studies, scientists learned that levodopa, which is a precursor of dopamine, can cross the blood-brain barrier.[113] The levodopa then converts into dopamine in the brain. Although the drug ameliorates Parkinson's symptoms, it does not prevent the substantia nigra nerves from dying, nor does it lead to the restoration of dopamine. It also loses effectiveness with time.

Recently, researchers discovered through autopsies that the progressive degeneration of the dopamine-3 (D3) receptor in Parkinson's disease is what causes the gradual loss of efficacy of levodopa.[114] Levodopa therapy is strengthened when patients concurrently take dopamine agonists, such as ropinirole. These stimulate dopamine receptors.[115,116] Added to this, scientists found a substance that slows dopamine metabolism, making more available to control Parkinson's symptoms. A monoamine oxidase B inhibitor, a medication developed to treat depression called Selegiline (deprenyl), had this effect. Drugs of this type can have uncomfortable side effects. Recently, the FDA decided against recalling one, Tasmar, even after three deaths from liver failure, because there are so few drugs available to combat Parkinson's.[117] But the agency did demand that the drug be relabeled.

Amantadine is an antiviral medication that was by chance found to relieve the symptoms of PD. When it was given to patients suffering from viral infections who also had PD, physicians noticed that the PD symptoms improved. Amantadine is not as strong as levodopa (L-dopa) but can be used to decrease the amount of L-dopa needed or can be used by itself in patients with mild PD.

Serendipity continues to play a role in discovering new treatments for PD. Former British soldier and stuntman Tim Lawrence accidentally discovered that ecstasy eases his PD. Ecstasy changes the levels of serotonin in the brain which may explain why Lawrence's condition improved. Because of the side effects from ecstasy, it will not become an acceptable treatment, but new drugs can be designed that resemble ecstasy without the side effects. This finding had been predicted by human studies performed by Dr. Brotchie. He discovered that levels of the 5HT-2c serotonin receptor in the human brain are different in patients with PD.

In the meantime, employing epidemiology, clinical studies, and *in vitro* research is, as always, invaluable in identifying the susceptible population and the plausible causes of Parkinson's. The disease seems to have both genetic and environmental causes. Some people are genetically disposed to Parkinson's. The higher occurrence of Parkinson's in rural areas suggests that pesticides and other contaminants that increase free radicals could activate Parkinson's. Yet, the ultimate cause of Parkinson's remains unknown.[118] As in Alzheimer's, both familial and sporadic forms exist. Researchers identified a discrepancy in the α-synuclein gene on chromosome 4q, responsible for the adult-onset familial form of Parkinson's. There is an overabundance of the α-synuclein protein in these patients.[119,120,121,122] Missense mutations, where the genetic code has additions or deletions that do not translate into a functional protein in the number 17 chromosome, have some responsibility for the defect in the α-synuclein gene.[123] The discovery of the α-synuclein has been called "the first major breakthrough in the understanding of the disease in thirty years."[124,125] The same combination of human-based research mapped juvenile-onset Parkinson's to chromosome 6q.[126] Another epidemiology and *in vitro* discovery found that genes on chromosome 2p13 may be responsible for the sporadic form of the disease.[127] A mutation of the parkin gene was identified in four brothers from an Arab family in northern Israel suffering from a rare form of PD, autosomal recessive juvenile. Researchers theorize that the parkin gene mutation may prevent the gene from being expressed and thus prevent the body from breaking down and removing some harmful proteins from the brain.

Scientists hope to elucidate exactly how the disease affects neurons but have not yet done so. The reason neurology is such a frustrating field is that we cannot answer what appear to be simple questions. *In vitro* research enhanced by computers and electron microscopy seems to lead in a promising direction, but it will take time. Early diagnosis of Parkinson's helps physicians treat symptoms more thoroughly. With the advent of the PET scanner, TMS, fMRI and other scanners, almost all components of the dopamine synapses can now be imaged in living humans. SPECT technology, mentioned in reference to Alzheimer's, is similarly useful.

Until the 1960s, neurosurgeons simply removed one or both of the thalami, the two-part regions of the brain through which sensory impulses travel. Thalamotomies were not that effective. The procedure, designed to mitigate trembling, is being replaced by high-frequency deep brain stimulation, using a wire inserted into part of the thalamus subthalamic nucleus or globus pallidus. Pallidotomy—the lesioning of the globus pallidus, another nearby brain sector, seems to improve function in Parkinson's patients, however there are risks.[128] Neurosurgeons are now evaluating transplantation and implantation procedures. Ten years after implantation, a graft of embryonic nigral cells in a man with Parkinson's survives and continues to release dopamine.[129] Stem cell research is also promising.

Epilepsy

Epilepsy, from the Greek word *epilambanein* meaning to attack or to seize, was first described in ancient times. Julius Caesar, Napoleon, and Van Gogh are among history's famous epileptics. Epileptic seizures are brief, recurrent attacks of altered consciousness and motor activity. During a seizure, epileptics' extremities often jerk violently, and sometimes victims lose consciousness. Alternatively, epileptics may display emotional outbursts or periods of confusion. Generalized seizures, seizures that involve the entire body are classified in two forms: tonic-clonic (which used to be called "grand mal") and absence seizures (which used to be called "petit mal"). Partial seizures are when just a portion of the body is affected.

With no explanation to the contrary, people of yore blamed this group of chronic neurological disorders on the supernatural and considered epileptics as "possessed." Consequently, victims of epilepsy have long been discriminated against. Only in 1982 did the state of Missouri repeal its law prohibiting people suffering from epilepsy to marry. Even as late as 1986, the state of South Carolina permitted forced sterilization of women with epilepsy.[130] Over two and a half million Americans are epileptics, and about 200,000 new cases are diagnosed annually. The financial cost is over $10 billion per year.

John Hughlings Jackson was one of the first moderns to describe epilepsy accurately. Where did he get his insight? From clinical observation. Long before EEGs (electroencephalograms), Jackson deduced that abnormal electrical discharges in the brain caused the seizures. During a seizure, brain waves register on EEGs with abnormal rhythm. Today, MRIs can often distinguish the seat of the disease and its underlying mechanisms. Researchers have identified many different types of epilepsy, all characterized by uncontrollable activity in the brain caused by excessive and synchronous nerve-cell discharges.

Computers help scientists chronicle epilepsy types. John G. R. Jeffreys stated, "The detailed models for focal interictal discharges arose largely from experiments on brain slices *in vitro,* combined with computer stimulations."[131] Some cases of epilepsy are inherited, but most are not. Defects in single genes or a combination of genes, exacerbated by environmental factors can lead to epilepsy. This we know from a combination of epidemiology and *in vitro* research. For example, isolated genetic defects are responsible for a rare condition known as tuberous sclerosis, which causes epilepsy.[132] This is but one of many examples. There is such variety in the causes of human epilepsy that the frequent use of nonhuman brains in research is just ridiculous. Why gather data about animal seizures when the human spectrum is so diverse *and* accessible?

As an example, take a look at one cause of epilepsy, cortical dysplasia, which is a result of brain damage that occurs *in utero.* (Cortical means "brain-related," and dysplasia has to do with abnormal cell growth.) The connection between this brain malformations and epilepsy was known in the 1800s, however no one could act on the knowledge until recently. Cortical dysplasia researcher N. Chevassus-au-Louis and his colleagues are in favor of the animal model. Yet, when they summarized the three most meaningful contributions to our understanding of epilepsy in the 1990s, they made no mention of animal models:

- Development and widespread use of modern brain-imaging techniques, specifically magnetic resonance imaging (MRI).
- The ability to detect malformations during life has opened the door for surgical resection of the malformed areas. Indeed surgery can reduce or abolish seizures or both in most patients, suggesting the existence of at least a causal link between malformation and epilepsy. Moreover, neuropathologic analysis of resected tissue has frequently demonstrated the presence of more subtle malformations that were not identified *in vivo.*
- Recent progress in human molecular genetics has allowed the identification of several genes whose mutations leads to both malformations and epilepsy.[133]

This is not to say that they or their colleagues refrain from using animal models. With the link and the therapy having been established in humans, researchers sought to copy the cortical dysplasia pathology in animals. They genetically altered animals while *in utero*, administering chemicals such as cocaine and alcohol to pregnant animals to create cortical malformations and seizures in their offspring. All of this was for naught. Certainly, plenty of babies born of substance-abused mothers demonstrate a problem. Nevertheless, scientist E. F. Sperber and colleagues stated in 1999 that the experimental animal models of cortical dysplasia did not mimic the clinical pathology seen in humans.[134]

Clinical studies found that infants' seizures are different from adults'. Infants have lower seizure thresholds than adults and respond differently than adults to antiepileptic drugs. Also, when seizures occur in infants they may create a predisposition to seizures or cognitive defects later in life. Animal models have reproduced these clinical observations. Various chemicals isolated in animal models may explain the phenomenon—but only in the animals studied. These findings have not influenced the treatment of humans whatsoever.[135]

Since many animals suffer naturally from seizures or can be made to seize from artificial causes, one can see why early experimenters thought they would divine something from animal models. But the long history of ineffectiveness has not borne out the logic of their decision. All told, animal models of epilepsy have many drawbacks, first in the inaccuracies of the way they mimic the disease, and second in their response to medications. Take human epilepsy caused by single gene defects. Regarding this, Jeffrey L. Noebels of Baylor Medical School wrote:

> Perhaps more surprising is the realization that within this first group of human epilepsy genes to be described, there is not one the role of which in neuronal synchronization had been previously implicated in experimental animal models of acute epileptiform seizures.[136]

There are six mouse models of single gene spike-wave epilepsy: tottering, lethargic, slow-wave epilepsy, stargazer, mocha, and ducky. Each rodent has seizures secondary to an absent gene. The problem is that, in each case, the absent gene also contributes to disorders that do not include epilepsy in humans. Also, the mice exhibit symptoms such as ataxia (uncontrolled movement) that humans suffering from single-gene epilepsy do not.[137] Thus, the humans missing the gene are very different from the mice missing the same gene.

The EL mouse and the audiogenic seizures (AGS) mouse model epilepsies are not of single gene spike-wave nature but are caused by multiple interacting factors. The area of the brain known as the hippocampus loses neurons in humans but not EL mice. Nor do EL mice exhibit hippocampal ganglioside loss evident in humans. This means that the underlying reason why these mice seize and the neurological results of the seizures controvert human data. Thomas N. Seyfried and his colleagues wrote, "The reason for the difference between man and mouse in seizure-associated neuronal loss is not clear, but it may reflect a species difference in either neuronal or glial response to seizures."[138] Yes, these mice seize. But the reason they seize and the impact of the seizure on the brain are nothing like the human experience. Models this disparate cannot be expected to yield useful data.

On the treatment side, we find much the same story: Human-based models generate useful drugs; animal models do not. As epilepsy is an ancient disease, the seeds of present-day treatments emerged serendipitously over the eons. Queen Victoria's physician, Sir Charles Locock, from his treatment of pregnant women, decided that hypersexuality, "impure thoughts," and masturbation must stimulate seizures. He introduced bromide as an antiaphrodisiac in 1857. It appeared to be effective. Locock was influenced in his choice of bromide by earlier events. Serendipity, as usual, played a key role. A chance observation in 1853 revealed that potassium bromide prevented a young woman from having further seizures. This paved the way for Locock to use the drug.

Then, in the early 1900s, Dr. Hauptmann gave barbiturates to mental asylum patients as a hypnotic and noticed that they too inhibited seizures. Phenobarbital was introduced as an antiepileptic drug (AED) in 1912 and is still used extensively.[139] Phenobarbital would not be approved today because it causes cancer in mice and rats.[140] Richard H. Mattson of Yale School of Medicine stated:

> Overwhelmingly, discovery of the old and a number of the new AEDs came from serendipity. For example, of the old drugs, phenobarbital was a weak hypnotic agent that by chance proved to be a good AED, and carbamazepine was developed to have characteristics similar to Thorazine yet was found to have strong antiepileptic capabilities. [141]

He continued:

> Valproate (VPA) perhaps best illustrates the element of chance in AED discovery. VPA was discovered and synthesized by Beverly Burton in 1882 as evidence in his thesis on propyl derivatives . . . the agent found its way to the chemist's shelf as a solvent. Almost eighty years later Meunier and Meunier discovered its anticonvulsant properties.[142]

Even though most sufferers can control their seizures with antiepileptic drugs, about thirty percent complain of severe side effects. One good reason for this is the predilection for developing drugs around animal models.

Once a drug has been shown to be effective against epilepsy in humans by means of serendipitous observation, an animal can usually be found in which that same drug will inhibit seizures. Other similar drugs are then tested on the animal to find what appears to be the optimal solution. This information can also be obtained from studying the structure of the drug instead. How many successful AEDs were lost because they did not effectively treat rats but would have been effective in humans? How many mediocre AEDs are on the market because they worked better on animals?

Consider dogs as an example. Naturally occurring epilepsy is relatively common in dogs, particularly certain purebreds, such as Labrador retrievers, but drug assimilation differs from humans in important ways. Firstly, dogs metabolize many medications much faster than humans do. Secondly, whereas the drugs may stop the seizures in both dogs and humans, they affect the dogs differently in other ways. For example, vigabatrin was effective in dogs but caused hemolytic anemia and visual field impairment in humans and had to be withdrawn.[143,144] A standard medication for epilepsy in humans, Dilantin, is not used regularly in pets. It is metabolized far too rapidly to be effective. Phenobarbital and potassium bromide continue to be the mainstays of treatment in dogs.

Cats, rats, and mice can also be made to seize but human treatment options developed along these lines have drawbacks similar to those in the dog. NMDLA-induced seizures in rats are commonly studied. W. Loscher stated about NMDLA-induced seizures:

> In other words, a potent new drug against NMDLA or NMDA-induced seizures is not necessarily a useful drug for therapy of drug-resistant epilepsy (in humans), as demonstrated by the disappointing data from clinical studies with NMDA antagonists in patients with refractory epilepsy.[145]

Felbamate was one of the first modern AEDs to be approved in the United States. It was thoroughly tested on animals and subsequently underwent an aggressive marketing campaign. *Time* magazine even ran an article titled "Taming the Brainstorms." Within months after its launch, in the early 1990s, reports of aplastic anemia began surfacing. Cases of liver failure and death ensued. Commenting on the side effects, physicians John Pellock and Martin Brodie stressed the exigency for more patient evaluations prior to giving drugs to the general public and thorough post-marketing drug surveillance after drugs are on the market. They did not call for more animal studies. Pellock and Brodie criticized the mass marketing associated with new drugs. They also recommended that Felbamate be used only in refractory cases.[146] Tragedies due to animal models are far bigger news than putative successes, but they get very little press. Animal models are wasting scarce dollars and leaving people sick. But news like that, as we have explained, does not sell.

Most successful AEDs are designed on the structure of a known AED or are novel drugs based on the shape of the receptor on the neuron surface. For instance, phenytoin or Dilantin, a popular AED, was tested on animals first, as vested-interest groups claim. However, its discovery owed nothing to animal models since it was an analog of Hauptmann's original phenobarbitol. It was the similarity in structure to a proven anti-seizure medication that led investigators to try it in the first place. Clinical observation had revealed that pheno-barbiturates had anti-seizure properties but other nonpheno versions of the barbiturates did not. This pheno structure was modified and phenytoin was the result.[147] Moreover, when it came time to screen the chemicals, the protocol that the investigators used to test the animals was later shown to be unreliable. When phenytoin was retested using the protocol without any errors, it did not elevate the animals' seizure threshold. In other words it was sheer luck that phenytoin worked in these animal experiments. The sloppy nature of phenytoin testing actually led scientists to question how many substances that could raise seizure threshold had been lost because of faulty animal studies.[148] Meijer and colleagues stated in 1983:

> Unfortunately many AEDs [antiepileptic drugs] show marked pharmacological differences between animals and man. . . . [149]

An important neurotransmitter of the central nervous system, GABA (gamma-amino-butyric acid), is known to inhibit excitatory responses. GABA was discovered in animal brains, but human brains from cadavers could have been used. Thanks to MRI spectroscopy, one of the more important epilepsy studies revealed that GABA levels are lower in human patients with seizures than in non-epileptics.[150] Many newer anticonvulsants are designed to raise GABA levels. Gabapentin, which researchers realized would inhibit the central nervous system in the 1950s, was initially developed to treat spasticity and then found to be effective against some seizures.

Nowhere has new technology influenced drug discovery as much as it has for AEDs. The investigation of specific molecular targets and subsequent development of drugs that interact with those targets has revolutionized AED discovery. Marvelously, *in vitro* techniques allow drugs to be evaluated by displacement of radioligands from a binding site or the inhibition of enzyme activity or neurotransmitter uptake. Computers conduct mass screening of large databases of chemicals looking for specific molecular configurations. Combinatorial chemistry generates a pool of likely chemicals with the same basic structure for testing. Alternatively, a known effective drug undergoes chemical modification and then computers and *in vitro* research evaluate its efficacy compared with the first drug. Even side effects and efficacy are apparent in this analysis. Drug discovery using this process is called *structure-activity analysis* or *structure-function analysis,* as we reviewed in the Development of Medications chapter.

The National Institutes of Health is using human brain tissue of epileptic patients, obtained during palliative surgeries for this disorder, to study why seizures occur and what medications can be used to treat them.[151] The director of a top epilepsy research facility in Europe said:

> As a scientist, I am of the opinion that animal experiments bring no progress in the diagnosis and therapy of epilepsies. I have a well-founded suspicion that similar facts apply in other areas of medicine.[152]

When the epilepsy cannot be controlled long term with medication the patient is said to suffer from intractable epilepsy (IE). Out of the approximately fifty million epileptics worldwide, fifteen million suffer from IE. Many research modalities are offering hope for these patients. Mathematical analysis of brain waves may enable physicians and patients to someday predict seizures and thus prevent them.[153]

Technology also improves the lifestyles of people with epilepsy.[154] Clinical studies revealed that brain signals change, indicating an inclement seizure, in advance, sometimes as long as one hour. By establishing a pattern for these changes, using a computer analysis of the EEG, physicians can help patients predict and plan for seizures. The technology may also locate the source of seizure in the brain, thus enabling neurosurgeons to remove it.

Psychiatric Medications

Psychiatric diseases are especially difficult to reproduce in animals. In any species, physical manifestations of emotions can signal any number of pathologies. For instance, hypoxic (short of oxygen) patients will experience palpitations. So do those who are having heart attacks or heart rhythm disorders or numerous other diseases. A higher pulse and respiratory rate indicate fearfulness, but they also augur fever, viral and bacterial infections, pain, and multiple other illnesses. All told, most psychiatric disorders can only be qualified and quantified through the patient's expression.

Without language, we can only conjecture about the source and extent of an animal's mental discomfort, not draw viable conclusions. *Agoraphobia* is anxiety about or avoidance of a place. Is an animal sick because it has an aversion to a veterinary clinic? *Anxiety disorder* is persistent and excessive worry, something hard to distinguish in an animal (especially in the canine toy breeds). *Dementia* involves impaired memory. How can you tell if a rat running around his cage is demented?

Schizophrenia manifests as speech disturbances, delusions, and hallucinations. How can we diagnose an animal as having these problems? How about *manic disorder?* Note that most Labradors suffer from persistently elevated and expansive moods. And certainly we have all seen our dogs look "depressed." It does not necessarily follow that they have a psychiatric condition caused by an imbalance of neurotransmitters in the brain. Dogs can suffer from chemical imbalances too, but when we say Fido is depressed we usually mean he is disappointed or sick, not mentally ill.

Long before we had medications for psychiatric illness, human patients were unceremoniously remanded to asylums and subjected to procedures such as lobotomy, exorcism, and tonics, and truly grueling conditions. Ancient Greeks treated women's *hysteria*—the source of the modern word—by fumigating the vagina because they believed that a tilted uterus was to blame. During the Middle Ages, simply being female increased chances of being chained in an asylum. Arbitrary mental-asylum confinements were questioned as in the 1700s, when Philippe Pinel of France advocated more humane treatment of the insane. Pinel's meticulous records clarified many mental disorders, but progress was dilatory.

Then in 1949, French anesthetist Henri Laborit remarked on the calming qualities of a newly conceived antihistamine, chlorpromazine. Laborit, like most anesthetists of the first half of the twentieth century, regularly sedated patients prior to and after operations with antihistamines. These are quite safe and cause very minimal respiratory depression. Laborit's chlorpromazine patients enjoyed, as he described it, a "euphoric quietude." As an experiment, he collaborated with two psychiatrists to give chlorpromazine to a psychiatric patient. Two years after being introduced, chlorpromazine had been prescribed to over two million patients and was already known as the "drug that emptied the state mental hospitals."[155]

At the same time, the research of two Indian physicians, Gananath Sen and Kartick Bose came to light. In 1931, they had published their findings about *Rauwolfia serpentina*, a snakeroot plant used medicinally in India for hundreds of years. Sen and Bose claimed that it calmed violently insane patients. By 1951, Swiss chemists had isolated the chemical responsible, reserpine, and marketed it as Serpasil, an early sedative. Unfortunately, Serpasil tended to induce depression too.[156]

In parallel, researchers began to understand that chemical disorders in the brain were causing anxiety, depression, schizophrenia, mania, and other mental illnesses. Though the malaise was "all in their head," the patients really were ill. This knowledge precipitated changes in mental health care. Slowly, medications to treat these chemical imbalances began revolutionizing psychiatry and allowing patients to lead normal lives.

We now know that the brain's limbic system (including the hippocampus, amygdala, and the ventral striatum) regulates the flow of information from sensory and motor pathways into consciousness. By controlling the interplay between the information using neurotransmitters, the brain modulates the emotions. Importantly, nerve cells in this area employ the neurotransmitter dopamine, a fact born out by genetic research on mental illness victims.

Overactivity in the nerve cells using dopamine—whether from high release of dopamine or from a surfeit of cell sensitivity—tends to induce psychosis. Chlorpromazine blocks dopamine receptors, just as reserpine does. It keeps patients from experiencing severe hallucinations, panic, and delusions, so effectively that it is still in use. Obviously, no animal tests predicted chlorpromazine's antipsychotic effect, just clinical observation, clear and simple. The medication was tested in animals after the clinical discovery and animals were found that would duplicate the knowledge already obtained from humans. As so often happens when a physician makes and reports a clinical discovery, animal experimenters repeated it and called it theirs. In this case, the historical reports speak for themselves.[157,158]

When dealing with chemicals that influence bodily functions, the structure of the chemical is very important. It is the structure that influences the function—in this case blocking dopamine receptors. Therefore, when a new medication appears there is a rush to synthesize other chemicals with a similar structure to determine if a better or similar medication can be made. Such was the case with chlorpromazine. Chlorpromazine led directly to the discovery of more psychiatric medications.[159]

The entire history of psychopharmaceutical development is one of refashioning chemicals around events discovered serendipitously. Enthusiasm for conjuring new drugs grew to a fever pitch around mid-twentieth century as scientists and pharmaceutical companies eagerly looked for "new" psychiatric drugs. Their investigation was random, not directed toward a particular substance; they were just trying to increase the number of weapons in the psychiatrist's arsenal. Meprobamate was introduced in 1955 as a centrally acting muscle relaxant, and subsequently found to have anti-anxiety properties. Remarketed, meprobamate became the first medication to be called an anxiolytic (providing anxiety relief).

Simultaneously, a chemist named Leo Sternbach was experimenting with various chemicals only desultorily. One day he was cleaning up his lab and came across some of the old chemicals still in their flasks. He noticed that one had crystallized. This caught his interest, so he sent it off to a pharmacologist for analysis. It had a unique chemical structure. Sternbach himself actually took the drug, which was a benzodiazepine, before any animal tests. It had a sedative effect. The benzodiazepines (drugs like Valium, Librium, Versed, and others) were the result.

The benzodiazepines were eventually tested on animals. Benzodiazepines placated aggressive animals (some of the time), but in humans they proved useful as anti-anxiety medications. There is no clear correlation between aggression in animals and anxiety in humans. Indeed, the two emotions are just that, two separate emotions.[160,161,162,163]

Also in the 1950s, tricyclic antidepressants (TCAs) appeared on the scene. Like chlorpromazine, they were originally designed to be antihistamines, but when tested on human patients they produced sedation.[164] Tricyclics were ignored for years, but with the chlorpromazine discovery that interest was renewed. The chemical structure of the TCAs and the benzodiazepines were similar in structure to chlorpromazine, so scientists thought maybe they also could be used to treat depression. The tricyclic imipramine, an iminodibenzyl tranquilizer, was introduced first and found to be effective in treating depressed patients.[165] Other tricyclics—amitriptyline, nortriptyline, and desipramine—followed. All these medications were synthesized because of their similarity to chlorpromazine. To claim they came from animal models is simply false.

The study of depression and anxiety in animals is perhaps the most perplexing type of research. Depression is alarmingly common in the United States with an estimated one in ten affected. It is hardly surprising that a cornucopia of grant money is available for researchers. What is unfortunate is that it is not used to study the animal that truly needs it—humans. Researcher F. G. Graef wrote that, ". . . the distance between animal models of psychopathology and modeled dysfunction is usually very large, the animal behaviour measured in laboratories being qualitatively different from the symptoms of the clinical condition."[166]

Cocaine, opiates and amphetamines would qualify as potential antidepressants as screened by many of our present animal models. For humans they are not satisfactory, although we now know that they act on several neuronal sites in common with modern antidepressants.[167]

As noted, the first successful treatments for depression materialized as they always do—by accident. During World War II, the Germans used hydrazine to power rockets launched against England and the rest of Europe. After the war, chemists were anxious to find cures for the tuberculosis (TB) that was then scavenging the world. They dipped into surplus hydrazine and derived iproniazid and isoniazid. Iproniazid was originally given to TB patients in hopes that it would help them clear their secretions. It did not. But the staff noticed that the patients receiving it were much happier than those not receiving it. Hence it came to be used as an antidepressant.[168] After iproniazid's properties had been demonstrated in humans, it was tested on animals. The results in animals were noncontributory to its use.

Pharmacologists learned that iproniazid and isoniazid inhibited mono-amine oxidase, an enzyme that normally breaks down norepinephrine. Too little norephinephrine can lead to depression, which is what was happening with reserpine, mentioned earlier. Mono-amine oxidase inhibitors work because they increase the presence of norepinephrine. The reserve of hydrazine resulted, serendipitously, in the discovery of mono-amine oxidase inhibitors (MAO-I), by now a psychopharmaceutical mainstay.

Reserpine, the snakeroot derivative, though causing depression, actually elucidated the pathophysiology of mental health. Reserpine was tried as a treatment for many diseases, high blood pressure being but one. Physicians noted that high blood pressure patients who received reserpine became depressed or had mood swings. When they stopped taking reserpine their moods returned to normal. This helped elucidate depression's biochemical basis. But when researchers successfully interrupted reserpine's depressive effects in animals using MAOI and tricyclic antidepressants, it created two monsters: one, the conclusion that depression was a simple matter of depleted monoamines and two, the reserpine animal model. According to scientists who wrestled with these "monsters,"

> Most of the approximately one hundred compounds that reached some stage of clinical trials in the last 25 years did so because they could "qualify" on these tests. Gamfexine was the first drug that failed to show a correlation between animal tests and human trials. Its effect on cats was exceptional but it worsened the clinical status of human patients, two of whom had to be prevented from committing suicide. Gamfexine was first in a long line of failures.[169]

MAOIs discovered clinically, like iproniazid, or synthesized to mimic iproniazid's effect, have worked out well. But contrast these victories with Meritol, which was released in 1986 after extensive animal testing. Meritol was promptly withdrawn secondary to severe life-threatening side effects including kidney failure, liver failure, hemolytic anemia, and deaths.[170] Mianserin was also introduced as an antidepressant after animal testing. Unfortunately, it caused severe problems with patients' blood. Ironically, the side effect could have been predicted by in vitro studies of human cells.[171,172,173] We emphasize, animal models did not work, but actually studying human cells would have. Why is this such a difficult reality to embrace? Clozapine is an antipsychotic medication that was introduced in 1980 and almost immediately withdrawn because it caused life-threatening blood disorders that were not predicted by animal models. It was reintroduced in the 1990s as a drug of last resort. Those taking Clozapine are required to take blood tests every few weeks. Of the setbacks, the scientists wrote:

Whether the "censorship" practiced by these animal models prevented us from developing chemically novel antidepressants remains an unanswered question. We do know that in the last 25 years not a single compound has been discovered which is unequivocally better in clinical efficacy than the very first drug of this class. It was the very failure of these animal models in screening out ineffective compounds that ushered in the next stage of psychopharmacologic research.[174]

By the 1970s, it was clear that the effectiveness of MAO inhibitors and tricyclics pivoted around their effect on the neurotransmitter serotonin, as clinical evidence linked serotonin to depression.[175,176] Autopsies on people who committed suicide showed lower levels of serotonin in the brain than in people who died of other causes.[177,178,179] The concentration of a serotonin metabolite, 5-HIAA, was lower in the cerebral spinal fluid of depressed patients versus nondepressed.[180,181,182] Subsequently, treatment with medications that raised levels of serotonin showed antidepressant effects.[183,184,185,186,187] It was then left for *in vitro* methods to find and test a chemical that would make more serotonin available to the nerves that influence depression.

Such work resulted in Prozac. Though the selective serotonin reuptake inhibitor, zimeldine was actually first discovered in 1971, Prozac (fluoxetine) is the first efficacious SSRI (selective serotonin reuptake inhibitor). Zimeldine had tested safely in animals but caused paralysis in humans and was withdrawn from the market. Ironically, Prozac, which is relatively harmless to humans, causes high blood pressure and increased heart rates in rats.[188] It also affects rats differently from humans in other ways. Prozac does not affect the cyclic-AMP response to norepinephrine challenge in rat forebrains, even after chronic administration.[189] These disparities underscore the uselessness of rats as models for human mental health.

After much animal experimentation in mental health research, we now know that the difference between rat serotonin receptors and the human counterpart is only one amino acid. The amino acid threonine replaces asparagine at place 355 in humans. However, this very tiny difference transforms the way drugs bind to two receptors.[190] The rat data was ignored and Prozac went to market and became the number-one-selling medication in the United States. Other SSRIs now include paroxetine, sertraline, and fluvoxamine. The success of these drugs has helped many to accept that physiological imbalances cause depression, and has removed the stigma from the condition.

Not all antidepressants work on all depressed people. For instance distinct brain differences, revealed in brain scans, showed that although Prozac reached the right brain areas, the shift in brain metabolism never occurred in some people.[191] This is important, not just to the afflicted, but also to our argument. If subtleties like this affect organisms within one species—humans—why would anything we can borrow from animal models be substantive?

Many scientists have recognized and spoken out on the issue of animal testing of psychiatric medications:

> Two major points emerge from our reading; the surprisingly poor track record of most if not all animal models to date (a) in accurately predicting clinically effective antidepressants and (b) in generating new and conceptually liberating hypotheses of the pathophysiology of depression. These observations are highlighted by the fact that almost every significant advance in antidepressant drug treatment from the discovery of iproniazid and imipramine to the recently introduced "second generation" class of antidepressants has resulted either from astute clinical observations or serendipity; a far cry from a planned, predictive, screening test. In fact many second generation antidepressants such as iprindol, mianserin, trazodone and salbutamol should be classified as false negatives on the conventional drug screening models (i.e., ineffective during preclinical screening but clinically efficacious). Conversely, a series of compounds, predicted to be at least as effective as imipramine, were reported to be clinically ineffective (i.e., false positives).[192]

An estimated twenty-percent of the nation's depression sufferers cannot be helped by the SSRIs like Prozac and Zoloft. Although these drugs may elevate their mood, the side effects make them intolerable. These downsides have scientists experimenting with blocking receptors for a small protein called Substance P, a neuropeptide that helps nerve cells communicate. Since 1931 researchers have struggled to understand Substance P in animals' brains. Seventy years later, they are finally working in the human arena.

There is more to treating mental illness than just drugs, clinical experience has taught. One form of depression, seasonal affective disorder (SAD), is successfully treated with light therapy. In the winter, some people are affected adversely by the shorter amount of daylight and thus become depressed. By adding more light to their environment, the depression is lessened or resolved. The exact mechanism of how the light resolves the depression is unknown, but clinical research has shown that it does work.

Anomalies in research for other psychiatric conditions reinforce the ineptitude of animal models. Scientists have attempted to learn more about panic attacks, for example, by making mice panicky. Consider the following definition from the *DSM-IV*, which is used by psychiatrists to describe mental-health disorders:

> A panic attack is . . . the sudden onset of intense apprehension, fearfulness, or terror, often associated with feelings of impending

doom. During these attacks, symptoms such as shortness of breath, palpitations, chest pain or discomfort, choking or smothering sensations, and fear of "going crazy" or losing control are present.[193]

Now imagine looking for these symptoms in a caged mouse.

Patients with anxiety disorders underwent MRI and SPECT scans and the scans showed an unusual distribution of benzodiazepine receptors in the brain. Using PET scans, scientists found that people with bipolar (manic-depressive) disorder have thirty percent more of an important class of signal-sending brain cells.[194] These are but two examples out of many human studies that have yielded useful, unequivocal results.[195]

Should they look for signs of schizophrenia in mice to decipher schizophrenia? No. It was clinical experience and self-experimentation that disclosed schizophrenia as a disease of the brain and not aberrant personality or the work of witches. A Swiss chemist Albert Hofmann was working with the fungus ergot when, after a long day, he felt unusual and dizzy. He went home and experienced hallucinations similar to those experienced by schizophrenics. He went back to lab the next day and purposefully took the chemical derivative of ergot that he had been working with the day before—lysergic acid diethylamide, which we call LSD. Hofmann's second "trip" was the same. His serendipitous finding suggested that schizophrenics might have too much of a chemical similar to LSD that cause their disturbances.

Another breakthrough in schizophrenic research arose from observing people who chronically abuse amphetamines. Physicians noticed that these people began to behave as if they were suffering from schizophrenia also. It was known that amphetamines enlarge amounts of dopamine released in the brain. This led to the hypothesis that schizophrenia is caused by abnormally high releases of dopamine at the synapses, or an increase in the sensitivity of cells to dopamine. Observing that when patients with Parkinson's disease get too much dopamine, they too develop symptoms of schizophrenia reinforced this hypothesis. Again, patients suffering from high blood pressure who were taking reserpine offered more clues. In those who were also schizophrenics, their schizophrenic symptoms improved when reserpine was begun. All human-based observations helped establish the biochemical basis of this mental illness.

Genetic research has made diagnosing psychiatric illness easier, as it appears to run in families. Gene research has actually identified one cause of schizophrenia as a defect on chromosome 6.[196] Comparisons of PET scans conducted while schizophrenics were hallucinating as well as while they were not allowed researchers to see which parts of the brain light up during psychotic breaks.

According to the results of a ten-year study by a team from Royal Ottawa Hospital, in Ontario, Canada, depressed people with a mutation in the gene encoding for a serotonin 2A receptor were more than twice as likely to commit suicide than those without the mutation. By identifying a biological reason for mental illness, researchers are optimistic that genetic tests and treatment will preempt the suicides. The team is now looking at whether people suffering from other mental disorders, such as schizophrenia, also carry the mutation.[197]

In another example of gene clues, variants in dopamine receptor gene D4 and in the dopamine transporter gene DAT1 are common among children with ADHD (Attention-Deficit Hyperactivity Disorder). Scientists feel these variants may limit their ability to regulate impulsive behaviors.[198]

Using brain scans, scientists found that introverts have more blood flow and activity in brain areas known as the frontal lobes and in the anterior thalamus, which are believed to be responsible for remembering, problem solving, and planning. In contrast, extroverts exhibit more activity in the anterior cingulate gyrus, temporal lobes and posterior thalamus—areas considered more involved in sensory processing such as listening, watching or driving.[199] Considering these fine distinctions in personality strictly within the human species, does it not seem probable that mental disorders are governed by equally subtle differences?

The animal model can otherwise botch medications too, with psychiatric ramifications. Many medications were tested on animals without apparent side effects, only to cause severe psychiatric disturbances in humans. Hallucinations occurred in patients given acyclovir, amphetamines, anticholinergics, antidepressants, antihistamines, barbiturates, benzodiazepines, isoniazid, ketamine, levodopa, methylphenidate, pergolide, and many others medications. Psychosis was seen to develop in patients taking steroids, anticonvulsants, bupropion, clozapine, cycloserine, quinidine, trimethoprim-sulfamethoxazole, and many others. Depression was found in patients taking HMG-CoA reductase inhibitors, isotretinoin, mefloquine, vinblastine and many others. Patients taking estrogens, sumatriptan and other seemingly unrelated drugs experienced panic attacks. Procaine derivatives gave some people a feeling of impending death. Calcium channel-blockers resulted in some patients becoming depressed.[200] The February 13, 1998, issue of *The Medical Letter* listed over one hundred medications and classes of medications that provoke psychiatric disturbances in human patients. Animal testing did not and cannot predict these things.

As seen, beneficial research, which has actually helped people with mental illness, historically issues from nonanimal based research. As a direct result of the ineffectiveness of animal modeling of mental illness, clinical psychologists largely ignore studies on animals. A study in 1979 revealed that only 7.5 percent of the articles referenced by scientists contained experiments on animals.[201] A similar study in 1986 revealed that only 33 out of 4,425 (0.75 percent) references in the clinical psychology literature were animal studies. In reviewing the 1984 volume of the *Journal of Consulting and Clinical Psychology,* Kelly found only 0.3 percent of references were animal-based. In reviewing the 1984 volume of *Behavior Therapy,* a journal that one would expect to use animal data, Kelly found only two percent of references cited referred to animal models.[202] A study published in the November 1996 issue of *American Psychologist* reported that only 5.7 percent of clinical psychologists felt that completely banning animal experimentation would be detrimental to their practice.[203]

However, the funneling of grant monies earmarked for mental illness study into animal labs continues. Sometimes, the funds are not even used for mental illness. The National Alliance for the Mentally Ill reports that the National Institute of Mental Health (NIMH), the government body dedicated to studying diseases of the brain such as schizophrenia, anxiety, and depression, spent more money studying AIDS in 1997 than schizophrenia. NIMH dedicated only thirty-six percent of its funds to studying the most important mental health diseases. [204] It appears the mentally ill are not receiving a fair share of the research dollars. As one scientist sums up:

> Many of the psychotropic drugs were discovered by chance when they were administered for one indication and observed to be helpful for an entirely different condition. The history of the development of both the major antidepressants and the antipsychotic drugs points up the fact that major scientific discoveries can evolve as a consequence of clinical investigation, rather than deductions from basic animal [-modeled] research.[205]

chapter 8

Beyond the Animal Model

> Happy is the man that findeth wisdom and the man that getteth understanding. . . . She is more precious than rubies, and all the things thou canst desire are not to be compared unto her. Length of days in her right hand; and in her left riches and honour.
>
> —Proverbs 3:13–16

As we have seen over the previous seven chapters, the use of the animal model to gain knowledge about human disease fails from both theoretical *and* practical points of view since data that would be relevant for humans cannot be reliably extrapolated from animal studies. Certainly a case can be made that animal models have provided some useful data—during the time before Darwin, before molecular biology, before genetics and other knowledge, when animals appeared to be more similar than dissimilar to humans. However, those times are long past, and in light of current scientific knowledge and the development of highly sophisticated technology and research modalities, animal models have outlived their usefulness.

We hope that we have been successful in communicating several key concepts regarding the failure of the animal model as a scientific paradigm:

- The notion that animal models are useful in the study of human disease cannot meet the rigorous demarcation criteria set forth by science for the study of scientific theory, most notably in its inability to be predictive, falsifiable, and progressive.
- The use of animal models violates the principle upon which all of modern biology is based, which is evolution.
- There are a huge number of instances in which the attempt to extrapolate data from animal studies to human disease resulted in serious harm to patients and delayed medical progress.

The continued use of animal models as a scientific paradigm is not only highly ineffective; it presents a direct and indirect threat to public health. The question now becomes: Where do we go from here? The answer may surprise you: We're already there.

There is a wealth of reliable research modalities that are not reliant upon animal models and provide useful data that are helping to save human lives and contribute enormously to the body of medical knowledge. Some of these research modalities, such as autopsy studies, trace their roots to ancient times; others, such as pharmacogenetics are so new that they seem to be taken straight from a science fiction writer's imagination. Whether they have been in practice for centuries or emerged only in the past couple of years, these modalities are responsible for the medical miracles, and the high standard of medical care, our society enjoys today. If animal-model research were abandoned today, and the funds to support it were channeled into these research modalities, there is no telling the tremendous strides we could make against heart disease, cancer, diabetes, and other diseases that plague our society.

Some refer to these nonanimal modalities as "alternatives" to animal-model research; however, that is a misnomer. An "alternative" implies that the original modality is a viable one, and that the "alternative" simply offers a different route to the same destination. Additionally, "alternative" may connote a less-than-desirable methodology. For example, we take an alternative route home from work because traffic is congested or the roads are closed. Normally, though, the main route we take is better—that's why it's our main route. We have shown that animal-model research is *not* a viable route to the destination of treatments and cures for human disease. It is not the "main" route to the answers we are seeking, nor are the nonanimal methodologies a less desirable "alternative" to that main route.

Autopsy Studies

Autopsy studies, which involve the internal and external examination of cadavers, are a time-honored method for gaining critical knowledge about the nature and function of the human body. They have even led to several of the most significant medical discoveries in history. For example, it was, in part, William Harvey's autopsy studies that were responsible for the first accurate observation, in 1622, that blood circulated from the right heart through the lungs back to the left heart and into the arteries and veins.

In modern times, autopsies are performed for the purpose of confirming the diagnosis at time of death; additionally, autopsy studies can be used to track the incidence/prevalence of disease, monitor public health, and confirm or refute the effectiveness of a new drug or surgical technique. Autopsies have been responsible for advancing our knowledge of diabetes, hepatitis, appendicitis, rheumatic fever, typhoid fever, ulcerative colitis, congenital heart disease, hyperparathyroidism, and almost every other illness.[1]

Unfortunately, autopsy rates are at an all-time low. This is especially disturbing because forty percent of the time, the cause of death listed by the physician is proven wrong or incomplete at autopsy.[2] According to an article in the medical journal *Chest,* autopsies prove that twenty percent of the time, the cause of death in the medical intensive-care unit at a tertiary-care hospital is different from what the physicians thought caused the death. The physicians in the medical intensive care unit are some of the best diagnosticians in medicine, and tertiary-care hospitals traditionally have offered excellent medical care. Of greater concern is the fact that this study reveals that if the correct diagnosis had been applied, roughly half of the patients would have received different treatments and may have lived.[3] All this points to the power of autopsy studies to reveal critical information about the human body—information that can advance medical knowledge significantly and save lives.

Clinical Research

Clinical research and observation—research ethically conducted with human subjects—is another time-honored method of scientific investigation. The first clinical researcher was Hippocrates, who fathered the concept in the fourth century B.C.E. Hippocrates believed that by observing enough cases, physicians could predict the course of a disease, both in terms of its likely effect and who was most at risk for developing it. Over the years, clinical research and observation has been vital to the discoveries that have earned Nobel Prizes. Today, it remains the ultimate arbiter of determining what is and is not useful to humans.

Clinical research and observation has given us the heart-lung machine, multiple surgical techniques, and our current approach to treating AIDS. It made possible our understanding of the differentiation between Hepatitis A and B. In addition, it suggested that wounds could be closed with sutures and that the pancreas had a role in diabetes. Studying humans identified genes that allow rapid progression from HIV to AIDS and different genes in long-term nonprogressors. It is also the final arbiter of a drugs efficacy and safety. These examples are far from exhaustive. Peruse any issue of the *New England Journal of Medicine* or *JAMA* and you will find most of the articles are from clinical research. Clinical research is why your physician treats you with the drug he does, performs the surgery she does or administers anything else in the fashion she does.

Epidemiology

Epidemiology is the field of research that studies the distribution of diseases in populations as well as the factors that influence the occurrence of disease. Epidemiologists study epidemics, which are outbreaks of disease that attack a large number of people at about the same time, and endemic diseases, which are diseases that exist permanently in a region.

The field of epidemiology is based on the premise that most diseases, rather than occurring randomly, are related to environmental and personal characteristics that vary by place, time, and subgroup of the population. Epidemiologists study the environmental and personal characteristics of people to determine how these impact on disease occurrence. These studies can yield a tremendous amount of information, including:

- Determining who is most likely to develop a particular disease.
- When the disease is most likely to occur.
- The pattern of disease occurrence over time.
- What type of exposure its victims have in common.
- How much exposure increases the rate of disease occurrence.
- How many cases of the disease could be prevented by eliminating the exposure.

It was epidemiology that provided the link between cholesterol and heart disease, smoking and heart disease and cancer, high blood pressure and stroke, diet and cancer and heart disease, environmental factors and cancer, and folic acid deficiency and spina bifida. Epidemiological studies have exposed environmental poisons, revealed causes of birth defects, and led to the abandonment of the practice of bloodletting. In addition, epidemiology has also shown how communicable diseases such as cholera and measles are transmitted, as well as how the HIV virus, which causes AIDS, is spread.

In Vitro Research

In vitro research is another nonanimal research modality that continues to advance our medical knowledge. Sophisticated cell and tissue preservation technology enables researchers to grow human cells and tissues outside the body in a controlled environment. The cells or tissue, which have been removed from humans during surgeries, biopsies, and autopsies, are cultivated in flasks, test tubes, or other special containers, enabling researchers to study the effects of a chemical substance on a living cell.

In light of the fact that most illnesses occur at the cellular and sub-cellular level, the importance of being able to explore the nature of disease exactly where it occurs—which is what *in vitro* research does—cannot be underestimated. *In vitro* technology is also invaluable for measuring the toxicity of chemicals. The National Institute of Environmental Health Sciences has created a center in North Carolina to study how chemicals turn "off" or "on" thousands of different genes clustered on a laboratory slide. These changes will then be read and analyzed using computer technology.

The Human Genome and Proteome Projects

The Humane Genome Project is an international scientific program, which in 2001 completed the process of analyzing the complete set of chemical instructions—the *genome*—that controls heredity in human beings. Scientists working on the project determined the exact order in which the bases—the four types of simple molecules that comprise DNA—occur.

Whereas bases can combine chemically in any order, they occur in almost the same order in all people. By identifying the order of bases in the human genome, scientists hope to develop individual, highly detailed *gene maps*. These maps will show the location of each gene, which carries the instructions for a particular characteristic, such as eye or skin color, on a chromosome. The map will also illuminate the specific gene's role in normal body processes or diseases.

The Human Genome Project (and the just-begun but even more complex and more important Human Proteome Project, which will categorize all human proteins) will open up exciting new vistas for disease cure and prevention by speeding up the identification of specific genes involved in disease. Considering that the study of humans revealed the genetic cause of such diseases as cystic fibrosis, sickle-cell anemia, testicular (and other) cancers, hereditary spastic paraplegia, familial hypertrophic cardiomyopathy, the information gained from the Human Genome Project represents the next logical step—or perhaps more accurately, a quantum leap—in the march of scientific progress.

Understanding the genome will affect science and medicine on numerous fronts. It will allow hundreds if not thousands of new drugs to be tailor-made for specific gene-induced diseases, such as heart disease and cancer, in specific individuals. By determining the structure of all human proteins, we will be able to design, on a computer, a drug that interacts with the protein as we wish. Further research on common genetic variants and the genes involved with all common diseases will allow people to have their genetic profile determined from a single blood sample. They can then use this information, with the guidance of their physician, to practice preventive medicine and focus on those organs and systems that are prone to dysfunction.

Individualized genetic profiles will also allow drugs to be tailor-made to treat a patient's disease and reveal the environmental factors to which a person is particularly sensitive. (For example, some people can smoke cigarettes for decades and not suffer from lung cancer, while others smoke for only a few years and die from lung cancer.) As research moves forward to perfect the process of gene therapy, the hope of changing faulty genes *in utero* or after a baby is born—long one of humanity's fondest dreams—can at last be achieved. Many single gene defects could be cured if we knew how to place a new gene safely into the human body.[4]

Perhaps the most exciting result of the Human Genome Project will be the effect it has on drug development, making genetic-medicine one of the most promising fields in medicine today.[5] Human-based research modalities such as receptor physiology, structure-activity relationship, physicochemical techniques, *in vitro* molecular toxicology, and high throughput screening have revolutionized drug development and discovery and new knowledge of the genome will complement these.

Pharmacogenetics is the field that studies the effect of genetics on individual variations in response to drugs. Pharmacogenomics is applied pharmacogenetics, a "gene-to-drug" strategy, enabling physicians to predict a person's response to a given drug before exposure to the drug. Pharmacogenomics promises to customize therapies to meet explicit genetic criteria. The Human Genome Project has already yielded knowledge of many genes and proteins such as *MPIF-1, KGF-2, VEGF-2, OPG* and TRAIL that may allow the manufacture of medications specifically designed to treat diseases of the bones and cardiovascular system, cancer, and other illnesses.[6] Greater understanding of how and why genes are expressed will lead to the development of drug therapy that turns off the process of cell replication in cancer.

An article in *Science* describes the enormous potential of the knowledge to be gained from the Human Genome Project:

> the Human Genome Project coupled with functional genomics and high-throughput screening methods is providing powerful new tools for elucidating polygenic components of human health and disease. . . . Automated systems are being developed to determine an individual's genotype for polymorphic genes that are known to be involved in the pathogenesis of their disease, in the metabolism and disposition of medications, and in the targets of drug therapy.[7]

As we understand more about how genes are involved in disease, we can test potential drugs to see what effect they have on the DNA that comprises the gene. Scientists can now impregnate a silicon chip of 2–3 cm width with strands of human DNA to observe directly if a particular drug triggers a gene to act, or stops it from acting.[8] The DNA chip also allows scientists to see how numerous cells will respond to the drug simultaneously.

With the completion of the Human Genome Project and the new Human Proteome Project, it is timely to contemplate the role of animals in the post-genomic age. More than 3,000 diseases from single-gene defects exist in humans, yet it is difficult to use animal models to identify human genes because an animal's genes, metabolism, and lifestyle are much different than a human's. Knowing this, researchers who refuse to abandon the animal model have come up with what they consider to be a solution: genetically engineered, or *transgenic* animals. Transgenic animals are artificially created species produced in the laboratory through the manipulation of an animal's genes. This is accomplished in several different ways, including cutting or recombining an animal's DNA, by adding or deleting segments of DNA, by injecting human genes into an animal, or by transferring genes from one species to another.

Transgenic animals have failed as models for human diseases for these reasons:

> Specific genetic defects can be as difficult to identify and characterize as those of their human counterparts; and affected animals often differ from unaffected controls in the genetic factors addition to the gene in question.[9]

Transgenic animals still have organs and systems that influence the intact organism apart from the gene in question, thus extrapolation ignores the differences.

Writing in the journal *ATLA*, T. Ben Mepham et al. explain why transgenic animals are nothing more than an empty promise:

> It is apparent from an analysis of some transgenic disease models that the *actual* benefits of using the models are rarely completely equivalent to the *potential* benefits . . . The currently available transgenic models for cystic fibrosis (CF) illustrate this point. None of the strains is ideal, with either the genotype and/or the phenotype of the mouse failing to accurately model the human condition. . . . There are several limitations in relation to the usefulness of the current approaches to developing transgenic disease models, particularly since many diseases are multifactorial. Problems persist when extrapolating data obtained by using such transgenic animals to the disease condition in humans.[10]

Even the pro-animal model literature admits to the ineffectiveness of transgenic animals. *Comparative Medicine* stated that animals might be used to find out which gene causes a disease in humans; however:

Results to date suggest that the predictive value of a candidate gene, established in such an animal model, is rather low. . . . In fact, it can be questioned whether the use of animal models is the most effective way to detect candidate genes for complex human disorders. Due to the complexity of the genotype-environment interactions, the pathways that lead to an aberrant phenotype often differ between man and animal.[11]

L.F.M. van Zutphen, Ph.D., of the department of Laboratory Animal Science, Utrecht University The Netherlands, states in *Comparative Medicine*:

The study of transgenic animal models is increasing our knowledge of gene function in physiologic and pathologic processes. However, the phenotypic effect of a transgene largely depends on the genetic background on which it is expressed, and is, therefore, still often unpredictable. This makes transgenesis often a rather inefficient procedure for creating animal models of human disorders.[12]

And the pro-animal experimentation textbook, Nonhuman Primates in Biomedical Research, Biology and Management stated:

Despite the enormous number of human diseases caused by single gene defects, a 1986 review of the literature revealed no instances in which research with nonhuman primates had revealed a well-defined hereditary disease controlled by a single gene.[13]

None of this should come as a surprise, given the fact that it is a denial of the fundamental principles of evolutionary biology, modern-day molecular biology, and genetics to continue using animals as models for humans.

Contributions from Other Scientific Fields

Research in the basic sciences, including physics, chemistry, mathematics, computer science, and statistics, has been vital to the advancement of medical knowledge and our current high standard of medical care. In addition to microscopes—perhaps the most important tool in all of science—physics has provided us:

- X-ray crystallography
- Synchrotron radiation sources
- Neutron sources
- Nuclear magnetic resonance and other spectroscopies for molecular structure determination

- Magnetic resonance imaging (MRI).
- Positron emission tomography (PET).
- Lasers, which have a wide range of applications in medicine, including eye surgery, the treatment of skin disorders, and as a replacement for the scalpel in many types of surgery
- Molecular tweezers
- Single spectroscopies

The field of chemistry has given us combinatorial and solid phase syntheses, DNA sequencing, biocompatible materials, drug delivery devices, polymerase chain reaction (to amplify DNA), separation and purification methods, and gene chips. Together, the fields of chemistry and physics have, through the power of technology, developed ways to use liquid oxygen, nitrogen, ether, eythyl chloride, Freon-12, and carbon dioxide in the treatment of oral and skin lesions, brain tumors, Parkinson's disease, and neuromuscular disorders.

Mathematics and computer science have given us Fast Fourier transforms used in spectroscopy and CAT scans, fast sequence alignment and database methods used in genomics, conformational search and optimization methods used in protein folding, and ecological and population models of disease.[14]

The field of statistics has made some of the most significant, yet least appreciated, contributions to the field of medicine. Though the field of statistics has been much maligned, without it there would be no way to analyze the enormous amounts of data that characterize the best epidemiological studies. In fact, the field of statistics has a long and distinguished history, dating back to the rise of probability theory in the 1600s with Blaise Pascal and Pierre de Fermat, which led to the development of modern-day statistics by Thomas Bayes and Karl Friedrich Gauss a century later. Great Britain led the way in statistics research, with Sir Ronald Fischer making numerous contributions, including the introduction of the concept of randomization and multivariate analysis. The use of statistics came of age with England's Sir Richard Doll study of smoking and British physicians in the 1950s. The NIH followed with similar studies in the 1960s. With the development of computer technology, the field of statistics has become an even more integral part of medical research.[15]

The Vital Role of Technology

All the nonanimal investigative modalities we've described, from epidemiology to clinical research, *in vitro* research, the Human Genome Project, mathematical modeling, and computer technology, represent the cutting edge of biomedical research. Today's most impressive discoveries—as well as those on the horizon—will continue to emerge from these arenas. But of all of these methodologies, the power of engineering and technology to change the face of medicine is the most inspiring.

For most people, the word *technology* brings to mind many of the accouterments of modern-day life—computers, the Internet, CD and DVD players, television, and the like. Yet, technology has a long history of creating ways to do things once thought impossible. For example, when the German physicist Hermann Helmholtz invented the ophthalmoscope in 1851, physicians were able to see inside the eye for the first time. The laryngoscope, which allows better visualization of the larynx, was invented in 1854. The concept of electrocautery, which is the use of electricity to coagulate bleeding vessels, came from Harvey Cushing, one of the world's greatest brain surgeons, in the early 1900s.

The microscope—one of the greatest tools in science—traces its roots to 1590, when Zacharias Janssen, a Dutch spectacle-maker, discovered the principle of the compound microscope. In the 1670s, the Dutch amateur scientist Anton von Leeuwenhoek made a single-lens microscope that could magnify up to 270x—and became the first person to observe microscopic life. Today, surgeons utilize stains and dyes to observe tissue under powerful microscopes that help them decide whether to remove tissue from a patient or treat more conservatively.

The general contributions of technology to the field of medicine could fill an entire volume (see the Surgery chapter for more details), but a few highlights of some of the more recent developments illustrate just how far science and technology can take us.

To see how far we have come through technology, one need only walk through today's technology-rich intensive care unit, where one-on-one nursing, aided by an array of monitors, are truly improving a patient's chance for life. Technology has given us blood-analysis equipment, heart catheters and angioplasty, arterial and venous catheters, lithotripsy (a noninvasive technique that uses shock waves to remove kidney stones), automatic internal defibrillators such as Vice President Dick Cheney received, X rays, CT scan, magnetic resonance spectroscopy (MRS), single photon emission computed tomography (SPECT), ultrasound, EEG, magnetoencephalography (MEGT), magnetic resonance imaging (MRI), functional MRI (fMRI), event related optical signals (EROS), and transcranial magnetic stimulation (TMS). All this highly sophisticated equipment is now providing physicians with a new level of information that they can use to ensure more accurate diagnoses of illness in a patient and an appropriate treatment program.

Patients with heart disease can now have their arteries examined with a noninvasive CT scan called HeartView. Advances in computer software and in CT scanners has made this device possible. The CT scan allows physicians to visualize soft plaque in addition to the more advanced hard plaque thus enabling them to start treatment or institute preventive measures sooner.

Tissue implants, including artificial eyes, heart valves, penile prosthesis, artificial blood vessels, pacemakers, and skin expanders used for harvesting more skin for skin grafts are all extending lives and improving quality of life. Today, we have computerized canes that operate in conjunction with sonar to help blind people avoid obstacles and walk without aid. Electronic hearing implants stimulate auditory nerves and help the deaf to hear.

Lasers have become one of the most important and versatile tools in surgery, where the laser beam's highly focused energy serves as a minute scalpel to remove birthmarks, minimize blood loss during surgery, and treat skin disorders. Ophthalmologists use lasers to reattach loose retinas and correct other potentially blinding conditions without anesthesia. Surgeons are also using lasers as intravascular scalpels to dissolve occlusions in patients experiencing a stroke. Dr. Wayne M. Clark of the Oregon Stroke Center and the Oregon Medical Laser Center in Portland, Oregon, have begun using infrared lasers to dissolve intracerebral clots during acute strokes. The results are preliminary, but encouraging.

Technology has also been the force behind several extraordinary developments in microsurgery. The microscope is used for the common procedure of placing tubes in a child's ears to decrease ear infections and hearing loss. It also allows microvascular surgeons to work in very small areas with minimal destruction of viable tissue, thus enabling them to reattach severed digits and limbs and save a part of the larynx in patients with advanced head and neck cancer. Surgeons can now remove herniated discs in the back in a procedure called microscopic discectomy. The microscope is also used in neurosurgery for the removal of tumors without damaging delicate nerves.

The development of new technologies has enabled surgeons to perform a wide range of operations less invasively, thus avoiding the large incisions, risk of infection, and long recovery periods involved in major surgery. In a procedure called *laparoscopy*, surgeons insert a laparoscope—a metal tube with lenses, an illumination system, and a channel for surgical instruments—into the abdomen through a small opening in the abdominal wall. Today, laparoscopy is used to perform appendectomies, cholecystectomies (gall bladder removal), hysterectomies, and hernia repairs. In a similar fashion, endoscopy is now used to scrutinize the upper gastrointestinal tract for ulcers or cancer as well as the inner colon for early cancers. An endoscope is a hollow tube with a lens on one end and fiber optics along its inside walls that is inserted directly into the organ or cavity being examined. Prior to the development of endoscopy, patients had to swallow a chalk-like substance, or receive it via an enema, followed by X rays in an uncomfortable procedure that did not always provide adequate visualization.

A new surgical technique has improved the treatment of uterine-fibroid tumors. Painful and difficult to treat, fibroid tumors can be reduced through medication, but the growths tend to recur when the patient stops taking them. They can be surgically removed either by a myomectomy or hysterectomy. Now, however, a less radical technique, which blocks the blood vessels that supply the fibroids, has shown success. Tiny plastic particles, which have been injected into a catheter, find their way from the femoral artery to the uterine artery. This is a variation on an old idea, that of cutting off the circulation to an area.[16]

Exciting new surgical technologies are on the horizon as well. Cryoablation uses extreme cold to kill pain-causing nerves and cancerous tissue. The same technology can also be used to heat an area of the body above 60° Celsius (140°F), the temperature at which tissue is destroyed. If the precise area of the body could be isolated from all others, cancer tissue could be destroyed without harming other areas of the body, as happens with chemotherapy. Researchers anticipate that they can achieve 100 percent success in destroying gall bladder and kidney stones using a process known as time-reversed acoustics (TRA). In TRA, a sound wave is sent directly back to the source that emitted. With current technology, only about one out of every three ultrasound blasts hits a stone, so this technology can dramatically improve the outcome for patients with these painful conditions.

Advances in diagnostic techniques are another example of the profound effect technology has had—and will continue to have—on medicine. Spiral computed tomography (spiral CT) and autofluorescence bronchoscopy have enabled physicians to detect early lung lesions that cannot yet be seen with conventional imaging modalities. Spiral CT images the lungs so quickly that data is collected in a single breath hold. In autofluorescence bronchoscopy, the patient inhales a pigment that causes tissue to take on different colors depending on whether it is normal or malignant. Inhaling an aerosol form of the compound also allows up to 100 times more of it to be delivered right to the lung lesion.

Acoustic microscopy is the ability to view a sample with sound waves instead of light waves. It is an extension of the ultrasound. Utilizing frequencies 1000 times greater than conventional ultrasound, the acoustic microscope may someday enable researchers better resolution than the traditional light microscope. This new technology puts the acoustic microscope in a class of microscopes including the X-ray microscope, and electron microscope. Another advantage of the acoustic microscope is that it may obviate the need for staining tissue in order to visualize the different components.[17]

The scanned-probe microscope allows scientists to see molecular and atomic shapes, mechanical properties, magnetic properties, electrical properties and temperature variations of a substance. And they do so without destroying the substance being studied. This device is in the same category as the electron microscope and the technique known as X-ray diffraction but without some of the downsides of these techniques.

The scanning tunneling microscope can look at the structure of surfaces on an atom-by-atom basis. It can actually see things one-hundredth of an atom in size. It can also reveal the atomic structure of a substance without destroying it. Think what this could mean for the study of viruses and other disease causing organisms.

Frequently, modern scanning technology can elucidate a condition to such an extent that surgery is no longer required; scientists can develop medications or other therapies to cure what once required an operation. Examining tissue on a molecular level demands X-ray crystallography of the sort that James Watson and Francis Crick used to understand DNA. X rays are directed toward a crystalized sample of the molecule and the diffracted X rays are recorded on a photography plate. Even more impressive is the detail produced by radiation emitted by electrons in a magnetic field. (Synchrotron radiation sources or neutron sources generate the magnetic field.) Following, technicians interpret the photo and define the structure of the molecule. Drugs can then be designed to either aid or block the molecule from binding to receptor sites.

An Israeli biotechnology firm, Given Imaging Ltd., has invented a camera that fits inside a pill. As the pill moves through the patient's digestive tract, it takes two pictures per second. The photos are transmitted to a receiver the patient wears on his belt. They physician then views the images as a video. This technology allows the physician to view most of the small intestine; previously, only about one-third of the small intestine could be seen by endoscopy. Additional impressive developments include: new DNA chip technology that can be used to differentiate clinically distinct cancer types and discover clinically important tumor subtypes; earlier prostate-cancer detection through the introduction of the 7-MHz ultrasound probe that is placed next to the prostate via the rectum—an improvement over the traditional digital rectal examination; new MRI scanners that have also enhanced the ability to diagnose and stage prostate cancer noninvasively; lasers that diagnose stomach and intestinal cancers when placed on the end of an endoscope and passed into the gastrointestinal tract via the mouth or rectum. (Cancer cells reflect the laser light differently from normal cells.)

Technology has overcome a major drawback of biopsy methods for detecting cancer of the uterus. A simple ultrasound scan can now detect uterine cancers with 96 percent accuracy,[18] whereas traditional biopsy methods sometimes miss the cancer because it only samples one tissue.

The study of the human brain and diseases of the brain are also being improved dramatically through technology. The combination of magnetoencephalography with functional magnetic resonance imaging (fMRI) and structural MRI will enable researchers to learn more about schizophrenia, epilepsy, stroke, autism, and brain damage from chemotherapy in children through brain mapping. Magnetoencephalography provides millisecond time resolution and identifies the source of neural activity, while fMRI provides high resolution of functional areas of the brain.

Technology has profoundly influenced drug discovery, and nowhere is that more evident than in the area of anti-epileptic drugs (AED). The investigation of specific molecular targets and subsequent development of drugs that interact with those targets has revolutionized AED discovery.

In a process known as structure-activity analysis, computers are used to conduct mass screening of large databases of chemicals in search of specific molecular configurations. Combinatorial chemistry facilitates the generation of many chemicals with the same basic structure for testing. A known effective drug will undergo chemical modification, and then either computers or *in vitro* research will evaluate its efficacy and side effects with the first drug. While structure-activity analysis has been used for decades, advances in technology and chemistry over the past ten years have enhanced and sped up the process significantly, even to the point where computers can evaluate chemicals via robotics using miniature, mechanized methods. As a result, the pharmaceutical industry can now test more than 160,000 chemicals in a few months.

The Power of Prevention

Despite the tremendous contribution of technology to the standard of medical care our society enjoys today, prevention is the one area that has the greatest potential to avert health problems before they occur. Whereas it is appropriate to look to medical science for advances in the treatment and cure of disease, it is equally appropriate that we take personal responsibility for behavior that has been shown to have a profound effect on our health. These behaviors include cigarette smoking, dietary choices, exercise, and alcohol abuse.

Cancer and diseases of the cardiovascular system kill more Americans than all other diseases combined, and in the vast majority of these cases the disease was preventable. In fact, about two-thirds of all diseases can be prevented. According to the Food and Drug Administration, tobacco products are responsible for more than 400,000 deaths a year. Studies by the Centers for Disease Control and Prevention have linked a poor diet and lack of exercise to more than 300,000 deaths annually. And 100,000 deaths a year can be attributed to alcohol abuse.

There is a huge body of evidence showing that a balanced diet with less saturated fat and more fruits and vegetables can minimize the risk of many diseases, including heart disease, cancers of the breast, colon, and prostate, osteoarthritis, diabetes, and gallbladder disease. In fact, it is estimated that a healthy, balanced diet rich in fruits and vegetables, combined with a regular exercise program, could eventually reduce the overall cancer incidence by 30 to 40 percent. Lung cancer could be reduced by 90 to 95 percent with a combination of diet, lifestyle changes, and quitting the tobacco habit.

Clearly, the importance of preventive medicine cannot be underestimated, which is why the distribution of healthcare research dollars in this country—in which the funds allocated to education or prevention is a tiny fraction of the total NIH budget—is completely out of balance. If the funds currently allocated to animal-model research were instead dedicated to preventive education programs, we are convinced that there would be a dramatic reduction in preventable disease.

We strongly believe that if the medical profession could teach preventive medicine early enough in life, people would lead healthier lifestyles that would help keep chronic, degenerative diseases at bay and contribute to longer, happier, and more productive lives. Changing the lifestyles of a large, diverse society is not an easy task, but with the proper resources it can be accomplished.

At the same time, it is important to recognize that not all diseases are preventable through proper attention to diet and lifestyle. Many people suffer from disease through absolutely no fault of their own, and to imply that changing a particular behavior, or set of behaviors, can cure all ills is unethical and irresponsible. Cystic fibrosis, sickle cell anemia, primary biliary cirrhosis, and a host of other diseases will never be cured by adopting a different lifestyle. But if those diseases that are preventable are given the right attention through educational programs, and are thus prevented from occurring, there would be no need for further research into those diseases. There would then be more funds available for research into diseases that are not presently preventable, and our chances of finding cures and treatments for those diseases would be greatly improved.

Changing the System

As we have seen, the contributions of nonanimal modalities to medical science are nothing less than awe-inspiring. And the potential of preventing or reducing many of our most feared diseases through changes in diet and lifestyle is tremendous. Imagine, then, how many lives would be saved, and what dramatic improvements in the quality of life could be provided for others, if the resources currently allocated to animal-model research were instead invested in these areas.

To accomplish that feat would mean changing the entire basis upon which funds for research are distributed. And, as we explored in Chapter 1, animal-model research is a deeply entrenched system that resists change on every level. Why? Because the status quo works in favor of animal researchers (and the industry built around them) in terms of monetary gain, career advancement, personal ego, and job security.

One congressman stated the way grant applications are funded is, "an old boy's system where program managers rely on trusted friends in the academic community to review their proposals. These friends recommend their friends as reviewers. . . . It is an incestuous 'buddy system.' "[19] Researchers become famous for conducting animal-model research and want to see it continue, since it justifies their reputations. As Lewis Wolpert states in *On Giants' Shoulders,* "I mean, what do we scientists thrive on? Not money—yes, of course we like money, but we thrive on the praise and admiration of our peers. That is our currency: praise and admiration."[20]

The cynic who quipped, "the rat is an animal which when injected, produces a paper" was right. Less than 15 percent of NIH grant applications are funded each year, and most of those are experiments involving animals. So, if you are a researcher at an academic institution that lives by the "publish or perish" maxim, then that is what you do. And for the purposes of getting a paper published, animal-model research is fast, neat, and tidy. You go home on Friday night and the animals will still be there in their cages when you return on Monday.

Human clinical research, though infinitely more valuable in terms of its ability to deliver applicable results, is much slower and more difficult. Judith Vaitukaitis said in *Clinical Research,* "Nothing is more demanding, more difficult, more frustrating, more time-consuming, and requiring more creativity than clinical research." Unlike animal-model researchers, clinicians have no control over their subjects, who may not return for follow-up appointments nor follow instructions. Human subjects may even be dishonest about their lifestyles. Animal researchers can addict monkeys to crack cocaine or heroin easily and safely in their nice, clean laboratories. But clinicians who want to study human crack or heroin addicts have little choice but to interact with some possibly unsavory characters.

Perhaps that is one reason why, in 1993, the National Cancer Advisory Board, declared that clinical research was in "crisis." Yet the next year, the National Cancer Institute (NCI), a division of NIH, allocated only one percent of its total R01 funds to clinical research.[21]

Animal-model research may be the easier way, but as we have amply demonstrated in both theory and application, it is not an effective way to find cures and treatments for human diseases. Under the current system, animal-model research protects only the vested-interest groups and does nothing to protect the greater interests of public health.

Is there a solution? Yes. We hope that by educating people about the pseudoscience of animal-model research, as well as how and why it is perpetuated, we can help bring about a groundswell of public pressure on lawmakers to reexamine the corrupt and wasteful system through which healthcare research funds are distributed. When more people begin to understand that human lives are lost every day because of scientists and corporate profiteers protecting their own interests, the seeds of change will be sown.

Postscript

James Madison said, "A people who mean to be their own government must arm themselves with the power which knowledge gives. A popular government without popular information or means of acquiring it is but a prologue to a farce or a tragedy or perhaps both." We hope, in this book, we have given people more power.

We have not, of course, addressed every issue in the animal-model debate. We refer the reader to the website of the National Anti-Vivisection Society (*www.navs.org*) for a detailed explanation of the ethical arguments and for many alternatives to using animals in dissection. The web sites of Americans For Medical Advancement (*www.curedisease.com*), Europeans For Medical Advancement (*www.curedisease.net*), and those of Niall Shanks (*www.etsu.edu/philos/faculty/shanks.htm*) and Hugh LaFollette (*www.etsu.edu/philos/faculty/hugh.htm*) have more information of a scientific nature. *Brute Science* (Routledge, 1997) and *Sacred Cows and Golden Geese: The Human Cost of Experiments on Animals* (Continuum International, 2000) also contain much material not presented here. For the vested-interest groups' viewpoint on animal models, we refer the reader to the web sites of Americans for Medical Progress (*www.ampef.org*) and the Foundation for Biomedical Research (*www.fbresearch.org*). We encourage the reader to compare the information on these sites.

In writing *Specious Science,* we have attempted to present the case that the results of animal models cannot be reliably extrapolated to humans in biomedical research, and therefore animal-model research is an outmoded, wasteful practice that should be replaced with a new paradigm based on modern evolutionary biology. Knowing that individuals on both sides of the issues—those who support animal-model research, and those who oppose it—will examine this work we would like to take this opportunity to address these two constituencies.

First, we would like those in the research community to know that this book, like *Sacred Cows and Golden Geese,* is for a segment of the general public. What we have tried to do is condense and simplify, as much as possible, the knowledge of someone with a graduate degree in science—someone who has also conducted a full literature review of all the specialties in medicine and medical history—and make it understandable to people with a minimal science background. We recognize that the explanations in this book are not as exhaustive as those that appear in the pages of *Nature, Science,* the *New England Journal of Medicine,* and other scientific publications; however, it was never our intent to adhere to the standards set by these publications in a book designed for the general public. If science journals such as the above would like us to defend our position in their pages, we are happy to comply with their standards.

We are frequently asked why we print statements from people in the animal-experimentation community condemning the animal model when we do not print their statements praising the animal model. The answer is simple. If someone has a vested interest in a product, statements in support of the product are plentiful; and meaningless. Just one leaked memo from the tobacco industry proved they knew nicotine was addictive, despite hundreds of statements to the contrary. Animal models don't predict human response. The fact that people who profit from the product endorse the product is meaningless but their unguarded statements condemning the product should be noted.

For those members of the scientific community who wish to debate us on a peer level, we welcome the chance to do so in an open forum, where we could discuss the animal-model paradigm in terms of epistemology, methodology, type I and type II errors, regulatory genes and evolution, signal transduction, G proteins, logic, vaccine development and testing, how the Nobel Prizes are won, and other related topics. Perhaps then these scientists could provide a list of well documented, predictive, reliable, prospectively studied, and currently used animal models—something which they have never been able to do. Considering all the times we have showed up for debates and the opposition has not, we doubt our calendar will fill any time soon.

Defenders of animal-model research are quick to offer historic examples of "breakthroughs" or speak in nebulous terms about the necessity of animal models. But we have to wonder, where are all the successful animal models that cannot be replaced? We have conducted an exhaustive search of the literature for several years, and have found little. So, with all due respect to the vested-interest groups in the scientific community, if drug testing can be carried out on animals with predictable results, please provide the protocols and prospective analysis performed at institutions without a vested interest in the outcome. If you say the alpha-omega test in rodents for carcinogenicity is reliable, then give us the data that prove it. Show how the animal model fared when a randomized group of chemicals was tested. (The NCI and other institutions have done this, and the animal models have been found lacking.)

We are frequently accused of taking the position that there are not and never have been any medical conditions in humans that can be replicated in animals. We do not understand why this criticism continues

to be leveled as we have repeatedly stated that animal models do occasionally get it right; one reason the germ theory of disease was accepted
was that animals died when injected with bacteria. We do not believe
that there are no medical conditions in which an animal could provide
an adequate research or clinical analogue to a human. We do believe
that because we cannot know which animal is like humans until we
know how humans react to the drug, surgery, and so forth, then animal
models are not predictive. And even though some may in fact sometimes
mimic the human condition, using them will result in more harm than
good. It is a gross misrepresentation of our statements and position to
charge us with believing that animals and humans have nothing in common or to accuse us of being special creationists.

In *Specious Science,* as well as in *Sacred Cows and Golden Geese,* we
have provided a wealth of examples where animal models provided false
or misleading data. It is not incumbent upon us to show that they got it
wrong every time. They did not and do not. *In order to prove that a
paradigm or model is not useful, one does not have to prove that it gets
the answer wrong every time. One need only show that it is unpredictable to the point of being right about the same percent of the time as
chance would dictate, or that it is wrong so often as to be unreliable.* In
light of the data we have presented, the animal-model researchers must
show huge numbers of times where animal models were correct *prospectively* in order to prove that the concept of using animals as models is
viable. A reliable paradigm or model should have a high ratio of rights
to wrongs. A ratio is composed of a numerator and a denominator. In
light of the huge denominator (wrongs) that we have presented here and
in *Sacred Cows and Golden Geese,* those who favor the animal model
must show an almost infinite numerator (successes) in order for the ratio
to be better than predicted by chance alone.

Finally, we would like to say yet again to our detractors: we do not
think human prisoners or any other humans should be experimented
upon against their will. Because we point out the great role clinical
observation and self-experimentation has played in medical discovery,
our detractors make the logistical leap that since we do not favor animal
experiments we must want to use humans. This is nonsense. Ethically
performed clinical research has played, is playing, and will continue to
play a vital role in medical advancement. To accuse us of wanting to use
humans as guinea pigs is typical of those who cannot counter our scientific arguments and thus must defame us in order to keep their position
in society.

Our examination of the animal model is based strictly on science, not
personal philosophy. Yet, the defenders of animal-model research, in
criticizing our view, accuse us of being biased against the methodology
because we love animals. We freely admit that we are animal lovers.
Also, we do not eat animals. None of this, however, has anything to do

with the validity of our argument against animal-model research. Science is not about a person's personal philosophy, religion, ethics, or morality; it is about the material world around us and how we can reliably and predictably describe that world. If Gandhi or Martin Luther King, Jr., were alive today and said that 2 + 2 = 5, they would be wrong. If Adolf Hitler or Pol Pot stated that DNA makes RNA makes protein, they would be correct. Albert Einstein had an illegitimate daughter whom he basically disowned, divorced his wife, and married his first cousin, yet he described the universe better than anyone had before. The concepts a scientist introduces should be judged on their ability to explain the material world, not on the basis of his or her personal beliefs or exploits.

While the animal-model industry has, not surprisingly, criticized our work, many in the animal-rights and animal-welfare movements have embraced it. We appreciate the members of the animal-protection community who have read our book and understand the principles upon which it is based. We urge them to evaluate all the information available on the issue, and to use such information appropriately and judiciously. There are a number of publications and groups that may seem to oppose the animal model, yet in reality they state unequivocally that it works and they therefore support it as a "necessary evil" until *alternatives* can be found. This type of thinking is just a guise for allowing the practice to continue. The vested-interest groups are even willing to call the animal model *evil* as long as the so-called animal-protection groups call it *necessary*. Necessary wins out every time. When deciding whether to support pro-animal organizations, see what their stance on animal experimentation is and then see what industries donate money to them. You might find a connection.

Some groups and publications agree with us that the animal model is scientifically untenable, but present the case against the animal-model using scientifically unsound principles. If you are an animal rightist or welfarist and want to use science as your ally in opposing the animal model, we advise you to use critical-thought processes based on science, reason, and logic to persuade people that their lives do not depend on the continuation of vivisection. People who base their position against vivisection on superstition, muddled thinking, or fallacious reasoning cannot hope to advance their cause significantly. It is also critically important to be consistent.

Inconsistency is the first thing people look at when evaluating a new idea. If one opposes animal models based on science, one had better know what science is and be consistent about it. For example, it is utterly inconsistent to oppose vivisection and then use the results of animal experiments to support other beliefs, such as the ill effects of genetically modified organisms. Animals are either scientifically tenable models for human physiology and disease, or they are not. You cannot have it both ways. If animals are unreliable models for human disease, then quoting a study that showed rats get cancer from a genetically modified food should be avoided.

During the French Revolution, the guillotine silenced Antoine Lavoisier, one of the greatest scientists of all time. When he was sentenced to death, he pleaded for more time, in order to finish one of his experiments. The judge stated, "The Revolution does not need scientists." It is disturbing to think that there are those in the animal-rights and animal-welfare communities who seem to agree with that statement. Those who disregard science will never be able to argue effectively against the animal-model industry, nor will they ever be able to persuade the general public that the animal model is not vital for tomorrow's breakthroughs. But those who understand science and use scientific facts and principles properly wield a far more effective and powerful weapon than emotion and rhetoric. And it is that weapon that will ultimately win the war against the use of the animal model in biomedical research.

We close this book with the quote from Bertrand Russell with which we begin it:

Ever since puberty, I have believed in the value of two things: kindness and clear thinking. At first these two remained more or less distinct; when I felt triumphant I believed most in clear thinking, and in the opposite mood I believed most in kindness. Gradually, the two have come more and more together in my feelings. I find that much unclear thought exists as an excuse for cruelty and that much cruelty is prompted by superstitious beliefs.[1]

References

Chapter 1: The Philosophy of Science

1. Russell, Bertrand. *The Autobiography of Bertrand Russell*. New York 1968: Atlantic Monthly Press.

2. Curd, Martin and Cover, J. A. *Philosophy of Science*. New York 1998: Norton, p. 211.

3. Ibid., p. 244.

4. Ibid., p. 212.

5. Ibid., p. 211.

6. Derry, Gregory N. *What Science Is and How It Works*. Princeton, NJ 1999: Princeton University Press, p. 202.

7. As quoted in LaFollette and Shanks. *Brute Science*. New York 1996: Routledge, p. 19. from Riordan, M., *The Hunting of the Quark*. New York 1987: Simon and Shuster, p. 136.

8. Curd and Cover, p. 144.

9. Ibid., p. 03.

10. Ibid., p. 7.

11. Ibid., pp. 27–37.

12. *AV Magazine* Fall 2000, p. 3.

13. This book is designed to provide a broad overview of the topic of animal-model research and related issues. For a more in-depth analysis of many of the subjects touched on here and in the following chapters, please refer to the Bibliography.

14. *New England Journal of Medicine* 2000; 342:42–49.

15. Greek, Ray and Greek, Jean Swingle. *Sacred Cows and Golden Geese: The Human Cost of Animal Experiments*, New York 2000: Continuum Publishing, p. 45.

16. Ibid., p. 73.

17. Ahrens, E. H. *The Crisis in Clinical Research: Overcoming Institutional Obstacles*. New York 1992: Oxford University Press.

18. *Human Epidemiology and Animal Laboratory Correlations in Chemical Carcinogenesis* Coulston and Shubick, eds. Ablex: 1980.

19. *Science* 1990; 250; 1331–2.

20. Ahrens.

21. *Science* 2001; 291:2547.

22. Committee on Addressing Career Paths for Clinical Research. Careers in Clinical Research: Obstacles and Opportunities. Institute of Medicine. Washington, D.C.: National Academy Press, 1994. Clinical Research Study Group: An analysis of the review (POR) of grant applications by the divisions of research grants, National Institutes of Health. Patient Oriented Research, Bethesda, MD, November 1994 NIH.

23. *Clinical Research* 1991; 39:145–156.

24. U.S. Congressional Hearings on Scientific Fraud and Misconduct. April 12, 1989.

25. *BMJ* 1988, Nov 5, 1151.

26. *Science*, 2000; 288:248–57.

27. *New England Journal of Medicine* 2000; 342:1645, *New Scientist*, June 10, 2000, p. 19.

28. International Studies in the Philosophy of Science (1994) pp. 195–210.

29. Wilson, J. G. [1978] "Feasibility and Design of Subhuman Primate Studies", in Wilson, J. G. and Fraser, F. C. (eds.) *Handbook of Teratology*, vol. 3, (New York: Plenum Press). pp. 260–261.

30. Park, Robert. *Voodoo Science: The Road From Foolishness to Fraud*. New York 2000: Oxford University Press, pp. 33–34.

31. Ibid., p. 26.

32. *The Flight from Science and Reason*. Eds. Gross, Paul, R, Norman Levitt, and Martin Lewis. The New York Academy of Sciences, 1996, p. 47.

33. Curd, Martin and Cover, J. A. *Philosophy of Science*. New York 1998: Norton, p. 27–37.

34. Modified from a table in *JAMA* 1998; 280:1618–19.

35. *New England Journal of Medicine,* June 8, 2000.

36. *Annals of Internal Medicine 2000;* 133:877–880, 911–913.

37. *New England Journal of Medicine 2000;* 343:1833–8.

38. Park, pp. 33–65.

39. Gross, Paul R. and Levitt, Norman. *Higher Superstition.* Johns Hopkins University Press, 1994, p. 24.

Chapter 2: The Theoretical Basis for the Failure of the Animal Model as a Scientific Paradigm

1. Primo Levi. The Periodic Table, Schocken. 1984 p. 60

2. Botting and Morrison in *Scientific American.* February 1997.

3. Trull, F. *Animal Models: Assessing the Scope of Their Use in Biomedical Research.* Charles River, MA. 1987: Charles River, pp. 327–36.

4. *Animal Research Fact vs. Myth.* Foundation for Biomedical Research.

5. The Foundation for Biomedical Research in *Animal Research and Human Health,* published by FBR 1992.

6. *AMA White Paper.* 1992.

7. Sigma Xi Statements of the Use of Animals in Research. *American Scientist,* 80, 73–76.

8. Garattini, S. and van Bekkum (Eds.). *The Importance of Animal Experiments for Safety and Biomedical Research.* Dordrecht 1990: Kluwer Academic Publishers, p. vii.

9. Michael S. Rand, DVM, Chief. Biotechnology Support Service, University of Arizona-Tucson, Lecture notes for September 8, 1999.

10. *Public Affairs Quarterly* 1993; 7: 113–30.

11. LaFollette, Hugh, and Shanks, Niall. *Brute Science.* New York 1996: Routledge, p. 63.

12. Ibid.

13. As quoted in LaFollette and Shanks. *Brute Science,* p. 101. From an article by King and Wilson that appeared in *Science* 1975; 188: 107–116.

14. As quoted in LaFollette and Shanks. *Brute Science.* Routledge 1996, p. 72.

15. Wolpert, Lewis. *The Triumph of the Embryo.* Oxford University Press 1991, p. 131.

16. As quoted in LaFollette and Shanks. *Brute Science.* Routledge 1996, p. 89. From an article by King and Wilson that appeared in *Science* 1975; 188: 107–116.

17. LaFollette and Shanks. p. 90.

18. Ibid., p. 77.

19. Ibid., p. 113.

20. Freedman, D. and Zeisel, H. in *Statistical Science* 1988; 3:1–2, 3–56.

21. Rand, Michael S., DVM, Chief. Biotechnology Support Service, University of Arizona, Tucson, Lecture Notes September 8, 1999.

22. Svendson, P. and Hau, J. *Handbook of Laboratory Animal Science.* Boca Raton, FL 1994: CRC Press.

23. LaFollette and Shanks.

24. As quoted in LaFollette and Shanks, p. 51. From an article by King and Wilson that appeared in Science 1975; 188: 107–116.

25. Elliot Paul. "Vivisection and the Emergence of Experimental Medicine in Nineteenth Century France." In Rupke, N. (Ed.) *Vivisection in Historical Perspective.* Croom Helm 1987, pp. 48–77. Bernard (1949), p. 1309. Originally published in 1865 Bernard, C. [1949]: *An Introduction to the Study of Experimental Medicine.* (Paris: Henry Schuman, Inc.).

26. Bross, Irwin. *Scientific Fraud vs. Scientific Truth.* 1991: Biomedical Technology Press.

27. As quoted in Bross, p. 107.

28. LaFollette and Shanks. Brute Science Routledge 1996, p. 52.

29. *New England Journal of Medicine,* 1999; 341:1653–60.

30. Curd, Martin and Cover, J. A. *Philosophy of Science.* New York 1998: Norton.

31. The workbook is published by the Massachusetts Society for Medical Research, an organization dedicated to promoting the animal model.

32. Park, Robert. *Voodoo Science: The Road From Foolishness to Fraud.* Oxford University Press 2000, p. 39.

1. *Nat Genet* 1999; 23:296–303.

2. *N Engl. J Med* 1999; 341:801–806.

3. *J Med Genet* 1999; 36:881–887.

4. Vandeberg, John L. *Nonhuman Primates in Biomedical Research. Biology and Management.* Bennett, Abee, Henrickson (Eds.), Academic Press, 1995, p. 140.

5. *ATLA* 1998; 26–27.

6. *The Scientist* March 19, 2001, p. 17.

7. *Science* 2001; 291:309.

8. *New Scientist* January 20, 2001, p. 3

9. *Science* 1999; 286:225–6.

10. *Science* 1999; 286:2244.

11. *Wall Street Journal,* October 11, 1999.

12. *Science* 1999; 286:2244.

13. *Science* 2000; 288:955.

14. *Nature* 2000; 403:820.

15. *The Scientist* February 7, 2000 p. 16.

16. *The Scientist* February 7, 2000 p. 17.

17. *Nature* 2000; 405:599.

18. *New Scientist* January 20, 2001 p. 8.

19. *The Scientist* February 7, 2000 p. 16.

20. *The Scientist* February 7, 2000 p. 16.

21. *NEJM,* 1998; 339:1186–93.

22. *New Engl J Med.* 2001; 344:175–181.

23. *N Engl Journal of Medicine/Medscape Wire,* January 30, 2001.

24. *Reuters Health,* June 29, 2000.

25. *Lancet,* 2000; 356:701–707.

26. *JAMA* 2001; 285:545–550.

27. Altman, Lawrence K. *Who Goes First? The Story of Self-Experimentation in Medicine,* University of California Press, 1998, pp. 256–273.

28. *JAMA* 2000; 283:589–90.

29. Altman, Lawrence K. *Who Goes First? The Story of Self-Experimentation in Medicine,* University of California Press, 1998, pp. 273–82.

30. Greenwald, Robert A. and Diamond, Herbert S. (Eds.) *CRC Handbook of Animal Models for the Rhematic Diseases, Vol. I.* CRC Press 1988, preface.

31. *Acta Allergol* 1965; 20:472–83.

32. *N Engl J Med* 1975; 293:1228–31.

33. *Folia Haematol Int. Mag Klin Morphol Blutforsch* 1965; 84:387–401.

34. Galli and Lantz, in Paul, William E. *Fundamental Immunology,* Lippincott-Raven 1999, 4th ed., p. 1130.

35. *Br Med J* 1972; 3:623.

36. *Haematologica* 1965; 20:472–83.

37. Altman, Lawrence K. *Who Goes First? The Story of Self-Experimentation in Medicine,* University of California Press, 1998, pp. 283–97.

38. *Proc Natl Acad Sci* USA 1999; 96: 12216–12218, 12810–12815.

39. *Nat Gent* 1999:23; 134–135, 166–175.

40. *Lancet* 1999; 354:2029–2031.

41. *Reuter's Health,* June 29, 2000.

42. Ron Winslow. *Drug Advances Bring New Hope to Cancer Battle. The Wall Street Journal,* May 14, 2001.

43. *Reuters Health.* December 8, 1999.

44. *Science* 1999; 286:531–537.

45. *Science* 1999; 286:531–537.

46. *N Engl J Med* 2001; 344:1031–1042, 1084–1086. ABC News, NBC News, and Reuters News Service, April 5, 2001.

47. As quoted on ABC News web site and ABC Radio News April 5, 2001.

48. Dr. Newlands, in *Successes, Failures and Hopes in Cancer Chemotherapy,* the proceedings of an international symposium, May 1978.

49. The Independent, p. 13 October 11, 2000.

50. *Nature* 2000; 408:307–10.

51. *Science* 1998; 282:284–7.

52. *J Natl Cancer Inst* 2000; 92:1260–1266.

53. *Annals of Internal Medicine* 1999; 131:247–255.

54. *Lancet* 1999; 354:975–978.

55. *N Engl J Med* 2001; 244:539–548.

56. From Medscape Pharmacotherapy, Pharmacogenomics, Genomics Approaches to Improving Drug Treatments. Medscape Pharmacotherapy, 2001. © 2001 Medscape, Inc..

57. *JAMA* 1999; 282:927–8 and 1801–2, 927.

58. *Science,* October 15, 1999, p. 444.

59. *Nature Genetics/*Medscape Wire, December 9, 1999.

60. *Reuters Health,* September 23, 1999.

61. *Obstetrics and Gynecology* 2000, 96: 65–69.

62. *CA* 1999; 49:327–8.

63. *Good Medicine,* from PCRM, Autumn 1999.

64. See references quoted in *Clinical Pharmacology and Therapeutics,* Vol. 7, No. 2, pp. 250–70.

65. Dr. David Conning, director general of the British Nutrition Foundation in *BNF Nutrition Bulletin* 1991; 16:37–44.

66. *Arch Intern Med* 1999; 159:2290–2296.

67. Altman, Lawrence K. *Who Goes First? The Story of Self-Experimentation in Medicine*, University of California Press, 1998, p. 283–97.

68. *Nature* 1999; 400:515.

69. Coulston and Shubic. (Eds.) *Human Epidemiology and Animal Laboratory Correlations in Chemical Carcinogenesis*, Ablex Pub 1980, p. 7.

70. Gross, Paul R. and Levitt, Norman. *Higher Superstition.* Johns Hopkins University Press, 1994.

71. As quoted in the *New York Times* December 6, 1993.

72. Coulston and Shubic. (Eds.) *Human Epidemiology and Animal Laboratory Correlations in Chemical Carcinogenesis*, Ablex Publishing 1980, p. 263.

73. *J Nat Cancer Inst* 1981; 6:1215.

74. Dr. Michael Utidjian of the Central Medical Department, American Cyanamid Company, New Jersey, USA, writing in the book *Perspectives in Basic and Applied Toxicology*, p. 309–329, ed. Bryan Ballantyne, publisher Butterworth, 1988.

75. *Science* Vol. 278, Nov. 7, 1997 p. 1041.

76. As quoted in *LA Times* Wednesday, May 6, 1998.

77. *Nature* 1999; 401:464–68.

78. *Lab Animal* June 2001, Volume 30, No. 6 issue, p. 13.

79. Beniashvili, Dzhemali Sh. *Experimental Tumors in Monkeys.* CRC Press 1994, p. 45.

80. Harrison, *Clinical Oncology*, 1980; 16:1–2.

81. *J Nat Cancer Inst* 1981; 6:1215.

82. *J Natl Cancer Inst* 1999; 91:1985–1986, 2009–2014.

83. *Nature*, July 29, 1999.

84. *Nature* 1999; 401:763–4.

85. *Kansas City Star*, November 11, 1999, Sec. B, p. 10.

86. Dr. Lester Lave, of the Carnegie-Mellon University in Pittsburgh, USA, and colleagues Drs. Ennever, Rosenkrantz, and Omenn, writing in the journal *Nature*, vol. 336, p. 631, 1988.

87. Philip H. Abelson, Deputy Editor, in his Editorial in *Science*, vol. 249, p. 1357, 1990.

88. Dr. C. Parkinson of the Centre for Medicines Research, Surrey, England, and Dr. P. Grasso of the Robens Institute of Health and Safety, Surrey, England, writing in the journal *Human and Experimental Toxicology*, 1993; 12: 99–109.

89. Ausman, Lynne. Associate Professor of Nutrition, Tufts. *Nutrition Reviews* 1993; 51:57–63.

90. *Am J Epidemiol* 1999; 150:869–877.

91. O'Keefe, S.J., Kidd, M., Espitalier-Noel, and G. Owira, P. *Rarity of colon cancer in Africans is associated with low animal product consumption, not fiber.* Am J Gastroenterol 1999; 94:1373–80.

92. *Good Medicine*, from PCRM, Autumn 1999.

93. *Int J Cancer* 2000; 85:73–77.

94. Gastroenterology 2000; 119.

95. *Nature Medicine* 1998; 4:1276–1280.

96. *Journal of the National Cancer Institute* 2001; 93:976.

97. *Cancer*, 1999:86:1675–1681.

98. *N Engl J Med* 1999; 341:2039–2048.

99. *JAMA* 1998; 280:1410–15.

100. Dr. A. J. Shorthouse and colleagues, writing in the *British Journal of Surgery*, 1908; 67:715–722.

101. Dr. Irwin Bross, e.g. in his US Congressional Testimony: *How We Lost the War Against Cancer*, 1981.

102. *J Clin Oncol* 1998; 16:1287–93.

103. *CAA Cancer Journal for Clinicians* 1999; 49 (No.5):258.

104. *Reuters Health,* October 13, 1999.

105. *Cancer* 1999; 86:1557–1566.

106. *Nat Genet.* 2001; 27(2):172–180.

107. *Cancer Res* 1999; 59:5160–5168.

108. *New Scientist,* September 11, 1999, p. 11.

109. *The American Journal of Pathology*, 1999; 155:739.

110. Boven, Epie and Winograd, Benjamin. (Eds.) *The Nude Mouse in Oncology Research.* CRC Press 1991 p. 278–9.

111. Boven, Epie and Winograd, Benjamin. (Eds.) *The Nude Mouse in Oncology Research.* CRC Press 1991 p. 292.

112. *Science* Vol. 278, November 7, 1997 p. 1041.

113. *N Engl J Med* 1999; 341:1565–71 and 1606–9.

114. *N Engl J Med* 2000; 342:42–49.

115. *J of Biological Chemistry*, Vol. 274, Issue 33, 23349–23357, August 13, 1999.

116. *Business Week*, November 22, 1999, p. 67.

117. *Reuters Health*, September 27, 1999.

118. *JAMA* 1999; 282:1801–2, and 927–8.

119. *Science*, October 15, 1999 p. 444.

120. *Reuters Health*, October 19, 1999.

Chapter 4: Development of Medications

1. *Reuters Health*, September 6, 2000.

2. *Science* Vol. 266, October 21, 1994 p. 368–9.

3. *Reuters Health*, October 12, 1999.

4. *NEJM* 1998 ;331:1085.

5. Stone, Trevor and Gail Darlington. *Pills, Potions, Poisons*, Oxford University Press, Oxford, 2000.

6. Lumley and Walker (Eds.). *Animal Toxicity Studies: Their Relevance for Man* Quay, 1990 p. 7.

7. *Principles of Animal Extrapolation.* Calabrese, Edward J. Lewis Publishing 1991 p. 225.

8. *Science* 1999;286:2361–2364.

9. *Nat Med* 2001;7:149, 167–73.

10. *Antimicro Agents Chemother* 1987; 31:1791–97.

11. *J Med Chem* 1996; 39:673–79.

12. *Pharmacotherapy* 1998;18:456–62.

13. *Antimicro Agents Chemother* 1998; 42:3251–55.

14. *Advances in Neurology* 1998; 76: 41–47.

15. *Nature Biotechnology* 1998; 16: 917–18.

16. *Journal of Biomolecular Screening* 1997; 2:249–59.

17. *Science* 1998; 282:396–9.

18. Norton, Ronald M., B.S. *Pharmacogenomics and Individualized Drug Therapy,* Medscape Pharmacotherapy, 2001.

19. Lazarou J, Pomeranz B. H., Corey P. N. *Incidence of adverse drug reactions in hospitalized patients.* A meta-analysis of prospective studies. *JAMA.* 1998;279:1200–1205.

20. *Science* 1999; 286:487–491.

21. *Science* 1999; 285:998–1002.

22. *Science* 1999;285:998–1002.

23. Ron Winslow. *Drug Advances Bring New Hope to Cancer Battle. The Wall Street Journal,* May 14, 2001.

24. *Nature Biotechnology* 1998;16: 887.

25. Kroetz, Deanna J., Ph.D. *Genomics and Proteomics in Drug Discovery and Development,* Millennial World Congress of Pharmaceutical Sciences San Francisco, California, April 16–20, 2000.

26. *Clinical Pharmacology and Therapeutics* Vol.7, No. 2 p. 250–70.

27. *N Engl J Med* 2000; 342:42–49.

28. Prof. D. R. Laurence, Professor of Pharmacology & Therapeutics at the School of Medicine, University College, London, and Dr. J. W. Black, Director of Therapeutic Research at the Wellcome Foundation Ltd, UK, *The Medicine You Take,* publishers Croom Helm, London, p. 108, 1978.

29. *Am J M Sc* 1943;206:642–52.

30. *Proc Soc Exper Biol & Med* 1956;91: 229–230.

31. *Lancet* 1962; 599–600.

32. *ATLA* 1994; 22:207–209.

33. *Principles of Animal Extrapolation.* Calabrese, Edward J. Lewis Publishing 1991 p. 203 and 404. Somers, G. F. in *Br J Pharmacol* 1960;15:111–116.

34. Somers, G. F. in *Br J Pharmacol* 1960; 15:111–116.

35. See Stephen, Trent and Brynner, Rock. *Dark Remedy: The Impact of Thalidomide and Its Revivial as a Vital Medicine.* Perseus 2001 and *Nature* 2001;410:411.

36. Roald Hoffmann. *The Same And Not The Same,* Columbia University Press 1995, p. 136.

37. *Exp Mol Path Supl,* 1963; 2:81–106.

38. *Federation Proceedings,* 1967; 26: 1131–6

39. *Teratogenesis, Carcinogenesis, and Mutagenesis* 1982 Vol.2, p. 361–74.

40. Prof. George Teeling-Smith, in A Question of Balance: The benefits and risks of pharmaceutical innovation, p. 29, publishers Office of Health Economics, 1980.

41. *The Scientist* January 22, 2001 p. 14.

42. Professor Andre McLean, Department of Clinical Pharmacology, University College MCM, London, speaking at a conference reported in *Animals and Alternatives in Toxicology,* p. 86, ed. M. Balls, J. Bridges and J. Southee, publisher Macmillan, 1991.

43. *Principles of Animal Extrapolation.* Calabrese, Edward J. Lewis Pub 1991 p. 174.

44. Dr. John Griffin, Director of the Association of British Pharmaceutical Industry, UK, quoted in the pharmaceutical magazine *Scrip*, p. 23, February 10, 1991.

45. Prof. D. Grahame-Smith, Department of Clinical Pharmacology, Oxford University, in *Risk-Benefit Analysis in Drug Research*, ed. Cavalla, 1981.

46. Dr. A. P. Fland, writing in *Journal of the Royal Society of Medicine*, Vol.71, pp. 693–696, 1978.

47. From Risk Assessment: Report of a Royal Society Study Group, publishers the Royal Society p. 72–73, 1983.

48. Professors Laurence, McLean & Weatherall, in the conclusion to their book *Safety Testing of New Drugs—Laboratory Predictions and Clinical Performance*, ed. DR Laurence, AEM McLean & M. Weatherall, publishers Academic Press, 1984.

49. *Toxicology In vitro*, August–October, 1999.

50. Lumley and Walker. (Eds). *Animal Toxicity Studies: Their Relevance for Man* Quay 1990 p. 73.

51. *Acute Toxicity Testing*, Academic Press 1998 p. 332.

52. Moore, Tina. *Deaths Halt Heart Drug Trials, Associated Press*, December 12, 2000.

53. *Science* 2000; 288:951.

54. Prof. R. W. Smithells, writing in the book *Monitoring for Drug Safety*, ed. Inman, pp. 306–313, 1980.

55. Lumley and Walker. (Eds.) *Animal Toxicity Studies: Their Relevance for Man* Quay 1990, p. 59.

56. *N Engl J Med* 2000;343:1826–32.

57. Novartis Pharmaceuticals Corporation. October 6, 2000.

58. *Dear Health Care Provider Letter*, November 1, 1999.

59. *Cephalalgia* 1999;19:651–654.

60. Willman, David. *How a New Policy Led to Seven Deadly Drugs, Los Angeles Times*, December 20, 2000.

61. *The Medical Letter* 2000;42:95.

62. Lumley and Walker (Eds). *Animal Toxicity Studies: Their Relevance for Man* Quay 1990 p. 59.

63. *N Engl J Med* 1975;292:599–603.

64. *The Journal of Immunology* 1965;95: 635–42.

65. Willman, David. *How a New Policy Led to Seven Deadly Drugs. Los Angeles Times*, December 20, 2000.

66. Willman, David. *How a New Policy Led to Seven Deadly Drugs. Los Angeles Times*, December 20, 2000.

67. Willman, David. *How a New Policy Led to Seven Deadly Drugs. Los Angeles Times*, December 20, 2000.

68. Willman, David. *How a New Policy Led to Seven Deadly Drugs. Los Angeles Times*, December 20, 2000.

69. Willman, David. *How a New Policy Led to Seven Deadly Drugs. Los Angeles Times*, December 20, 2000.

70. *Los Angeles Times*, Janet Lundblad.

71. Willman, David. *How a New Policy Led to Seven Deadly Drugs. Los Angeles Times*, December 20, 2000.

72. Willman, David. *How a New Policy Led to Seven Deadly Drugs. Los Angeles Times*, December 20, 2000.

73. Prof. George Teeling-Smith, in *A Question of Balance: The benefits and risks of pharmaceutical innovation*, p. 29, publisher Office of Health Economics, 1980.

74. The *LA Times* printed scores of articles, between 1998 and 2001 on the Rezulin affair from which this section is taken. If the reader wishes more information we suggest the *LA Times* web site, which is excellent.

75. Willman, David and Lundblad, Janet. *Risk Was Known as FDA Approved Fatal Drug, Los Angeles Times*, March 11, 2001.

76. Willman, David. *How a New Policy Led to Seven Deadly Drugs. Los Angeles Times*, December 20, 2000.

77. Dr. Anthony Dayan. Wellcome Research Laboratories, in *Risk-Benefit Analysis in Drug Research*, ed. Cavalla, p. 97, 1981.

78. Sir Douglas Black, Royal College of Physicians, in *Risk-Benefit Analysis in Drug Research*, ed. Cavalla, p. 187, 1981.

79. *Principles of Animal Extrapolation.* Calabrese, Edward J. Lewis Publishing 1991 p. 5–7.

80. *Reuters Health* February 8, 2001.

81. *JAMA* January 24, 2001.

82. Gad, Shayne C. and Chengelis, Christopher P. *Acute Toxicity Testing* Academic Press 1998 p. 332.

83. Caldwell, J. 1992: "Species Differences in Metabolism and their Toxicological Significance", *Toxicology Letters* 64 5, pp. 106.

84. Fox, Maggie. Health and Science Correspondent, *Reuters,* August 19, 1999.

85. *Principles of Animal Extrapolation.* Calabrese, Edward J. Lewis Publishing 1991 p. 206.

86. *Animal Models of Toxicology.* Gad, S. C. and Chengelis, C. P. (Eds.) Marcel Dekker Inc. 1992 p. 21–76.

87. *Principles of Animal Extrapolation.* Calabrese, Edward J. Lewis Publishing 1991 p. 45.

88. *Principles of Animal Extrapolation.* Calabrese, Edward J. Lewis Publishing 1991 p. 5.

89. *Principles of Animal Extrapolation.* Calabrese, Edward J. Lewis Publishing 1991 p. 90,102–3.

90. NAS 1977 in *Principles for Evaluating Chemicals in the Environment.* National Academy of Sciences, Washington DC pp. 156–97.

91. *Principles of Animal Extrapolation.* Calabrese, Edward J. Lewis Publishing 1991 p. 485.

92. *Principles of Animal Extrapolation.* Calabrese, Edward J. Lewis Publishing 1991 p. 203 and 404. Somers, G. F. in *Br J Pharmacol* 1960;15:111–116.

93. Somers, G. F. in *Br J Pharmacol* 1960;15:111–116

94. *Principles of Animal Extrapolation.* Calabrese, Edward J. Lewis Publishing 1991 p. 492.

95. *Neurotoxicology In vitro,* Pentreath, V. W. (Ed.) Taylor & Francis 1999, p. 21.

96. *JAVMA* 1987;191:1227–30.

97. *JAVMA* 1987;191:1227–30.

98. *Neurotoxicology In vitro,* Pentreath, V. W. (ed.) Taylor & Francis 1999 p. 9.

99. Dr. Ralph Heywood, director Huntington Research Center (UK) 1989

100. Lumley and Walker (eds). *Animal Toxicity Studies: Their Relevance for Man* Quay 1990 p. 57.

101. *Science* 1979; 204;587–5930.

102. *Nature Biotechnology* 1996;14: 1655–6.

103. *Nature Medicine* 1999;5:1110–12.

104. *Neurotoxicology In vitro,* Pentreath, V. W. (ed.) Taylor & Francis 1999.

105. *Toxicology In vitro,* August–October, 1999.

106. *Neurotoxicology In vitro,* Pentreath, V. W. (ed.) Taylor & Francis 1999 p. 151.

107. *NY Times,* "DNA Chip May Help Usher in a New Era of Product Testing," by Andrew Pollack, November 28, 2000.

108. *Science* 1998; 282:396–9.

109. *Neurotoxicology In vitro,* Pentreath, V. W. (ed.) Taylor & Francis 1999 p. 129.

110. *Neurotoxicology In vitro,* Pentreath, V. W. (ed.) Taylor & Francis 1999 p. 25–29.

111. *Neurotoxicology In vitro,* Pentreath, V. W. (ed.) Taylor & Francis 1999 p. 10.

112. Dr. Dennis Parke, Dept. of Biochemistry, University of Surrey, UK, writing in the journal *ATLA* Vol.22, p. 207–209, 1994.

113. *Nature Biotechnology,* 1996;14: 1655–6.

114. Pollack, Andrew. *New York Times,* November 28, 2000.

115. Dr. R. Brimblecombe, Vice-President of Research and Development, Smith, Kline and French Laboratories, in *Risk-Benefit Analysis in Drug Research,* Cavalla ed., p. 153, 1981.

116. Dr. Roy Goulding, clinical toxicologist, formerly head of the Poisons Unit at Guy's Hospital, London, in a speech given on 12th November 1990. P. Botham and I. Purchase in *News Scientist* May 2, 1992.

117. *JAMA,* 1999;282:1453–1457.

118. *JAMA* 1999;282:1453–1457.

119. *N Engl J Med* 2000; 343:1616–1626,1643–1649.

120. *N Engl J Med* 2000; 343:1616–1626,1643–1649.

121. Adams, Chris. *How Adroit Scientists Aid Biotech Companies with Taxpayer Money. The Wall Street Journal,* January 30, 2001.

122. Adams, Chris. *How Adroit Scientists Aid Biotech Companies with Taxpayer Money. Wall Street Journal,* January 30, 2001.

123. Adams, Chris. *How Adroit Scientists Aid Biotech Companies with Taxpayer Money. Wall Street Journal,* January 30, 2001.

124. *N Engl J Med* 2000;342:1645.

125. *New Scientist* June 10, 2000 p. 19.

126. *JAMA* 2001; 285:437–443.

1. Moneim A. Fadali, M.D., F.A.C.S. Cardiovascular and Thoracic Surgeon, UCLA, Los Angeles, California, in a video interview with *CIVIS* representative Kathy Ungar in March 1986.

2. Rene Dubos. Pulitzer Prize-winner and professor of microbiology at the Rockefeller Institute of New York, wrote in *Man, Medicine and Environment*. Praeger, New York, 1968, p. 107

3. Westaby, Stephen and Bosher, Cecil. *Landmarks in Cardiac Surgery,* ISIS Medical Media 1997, p. 224.

4. *Operations That Made History,* Greenwich Medical Media Press, 1996, pp. 17–25

5. Westaby, Stephen and Bosher, Cecil. *Landmarks in Cardiac Surgery,* ISIS Medical Media 1997, p. 227.

6. *Operations that Made History,* Greenwich medical Media Press, 1996, p. 7–16.

7. Walker K. *The Story of Medicine* Hutchison 1954.

8. *Companion Encyclopedia Of The History Of Medicine,* Bynum and Porter eds., Routledge 1993.

9. Asimov, I. *Asimov's Biographical Encyclopedia Of Science & Technology,* 2nd edition. Doubleday and Company 1982

10. Bayly, Beddow. *Clinical Medical Discoveries,* National Antivivisection Society, London 1961

11. Peller, S. *Quantitative Research in Human Biology and Medicine,* John Wright & Sons Ltd. 1967

12. Inglis, B. *A History of Medicine,* Wiedenfield & Nicholson, 1965

13. *Lancet,* November 9, 1957, p. 903–6

14. *A Century of Vivisection and Antivivisection,* Westacott, E., Daniel, C. W. Co. Ltd. 1949.

15. Sand, R. *The Advance to Social Medicine,* Staple Press 1952.

16. *Br Med J,* August 16, 1890 p. 377–9.

17. *The Advance to Social Medicine,* Staple 1952.

18. *Qualitative Research in Human Biology,* J. Wright and Sons, 1967

19. Valley-Radot, Rene. *The Life of Pasteur,* Garden City Publishing 1923.

20. *Lancet* October 25, 1958 p. 892.

21. Keys, Thomas E. *The History Of Surgical Anesthesia,* Wood Library-Museum Of Anesthesiology. 1996, p. 3.

22. Keys, Thomas E. *The History Of Surgical Anesthesia,* Wood Library-Museum Of Anesthesiology, 1996 p. 38.

23. *Animal Experimentation: Its Importance and Value to Human Medicine.* Board of Regents: American College of Surgeons Chicago 1933 p. 12.

24. *Anesthesia,* Fourth Edition, Churchill Livingstone, 1994.

25. Keys, Thomas E. *The History Of Surgical Anesthesia,* Wood Library-Museum Of Anesthesiology, 1996, p. 14.

26. Altman Lawrence K. *Who Goes First? The Story of Self-Experimentation in Medicine,* University of California Press, 1998, p. 56–7.

27. Davy, H. *Researches, Chiefly Concerning Nitrous Oxide.* Vol. III of *The Collected Works of Sir Humphrey Davy Elder* 1939, p. 271, 274; and Vol. IV of the same series *Relating to the Effected Produced by the Respiration of Nitrous Oxide Upon Different Individuals,* BMJ 1972; 2:367–68.

28. Davy, H. *Researches, Chemical and Philosophical, Chiefly Concerning Nitrous Oxide; or Dephlogisticated Nitrous Air, and Its Respiration,* J. Johnson 1800, p. 465, 556.

29. Keys, Thomas E. *The History Of Surgical Anesthesia,* Wood Library-Museum Of Anesthesiology, 1996 p. 15–18

30. Walker, K. *The Story of Medicine,* Hutchison 1954

31. B. Inglis. *A History of Medicine,* Wiedenfield and Nicholson, 1965

32. McGrew, R. *Encyclopedia of Medical History,* Macmillan 1985

33. Thompson, W.A.R. *Black's Medical Dictionary,* A & C Black 1981

34. Taylor, F. L. *Crawford W. Long and The Discovery of Ether Anesthesia,* Hoeber 1928, p. 63–4.

35. *Southern Medical Journal,* 1849; 5: 705–13

36. Walker, K. *The Story of Medicine.* Hutchison 1954)

37. *Lancet* 1937, October 23, p. 950

38. Keys, Thomas E. *The History Of Surgical Anesthesia,* Wood Library-Museum Of Anesthesiology, 1996 p. 32–33

39. Richardson, B. W. *Biological Experimentation,* 1896, p. 54

40. Wangensteen, O. H. *The Rise of Surgery: From Empiric Craft to Scientific Discipline,* University of Minnesota Press 1978

41. Richardson, R. G. *The Surgeons Heart: A History of Cardiac Surgery,* William Heinemann Medical Books Ltd.

42. *Goodman and Gilman's The Pharmacological Basis of Therapeutics,* Seventh edition, MacMillan Publishing 1985.

43. Keys, Thomas E. *The History Of Surgical Anesthesia,* Wood Library-Museum Of Anesthesiology, 1996 p. 39–41.

44. *Anesthesia,* Fourth Edition, Churchill Livingstone, 1994.

45. Altman, Laurence K. *Who Goes First? The Story of Self Experimentation.* Random House 1987.

46. Keys, Thomas E. *The History Of Surgical Anesthesia,* Wood Library-Museum Of Anesthesiology, 1996 p. xxii.

47. Altman, Lawrence K. *Who Goes First? The Story of Self-Experimentation in Medicine,* University of California Press, 1998, p. 71.

48. Sneader, W. *Drug Discovery; The Evolution of Modern Medicine.* Wiley 1985.

49. *BMJ* February 17, 1951, p. 353

50. Altman, Laurence K. *Who Goes First? The Story of Self Experimentation.* Random House 1987.

51. *Goodman and Gilman's The Pharmacological Basis of Therapeutics,* Seventh edition, MacMillan Publishing 1985.

52. *Medical World* May 12, 1939.

53. Svendsen, Per. Laboratory Animal Anesthesia in Svendsen, P. and Hau, J. (eds.) *Handbook of Laboratory Animal Science* Vol. I. CRC Press, p. 311–337.

54. Brodie, B. *Clinical Pharmacology and Therapeutics* 1962, Vol. 3, p. 374–80.

55. Svendsen, Per. Laboratory Animal Anesthesia in Svendsen, P. and Hau, J. (eds.) *Handbook of Laboratory Animal Science* Vol. I. CRC Press, p. 311–337.

56. *BJA* 1993 Vol. 71, pp. 885–94.

57. *Report of Royal Commission on Vivisection* 1912, p. 26.

58. Beddow Baly. *The Futility of Experimenting on Living Animals,* NEAVS, London 1962.

59. Tait, L. Letter, *Birmingham Daily Mail,* January 21, 1882.

60. Tait, L Trans. *Birmingham Phil. Soc.,* April 20, 1882.

61. Sir Frederick Treves *Br Med J,* 1898 2; pp. 1385–90.

62. Tait, L. Letter, *Birmingham Daily Mail,* January 21, 1882.

63. Hill, R. B. and Anderson, R. E. The Autopsy—Medical Practice and Public Policy, Butterworth 1988.

64. *British Journal of Surgery* 1949; 37: 92.

65. *Operations that Made History,* Greenwich Medical Media Press, 1996, p. 17, 38–41.

66. Sir Frederick Treves *Br Med J,* 1898 2; pp. 1385–90.

67. *Lancet* 1930, October 11, 784).

68. *Companion Encyclopedia Of The History Of Medicine,* Bynum and Porter eds., Routledge 1993.

69. Asimov, I. *Asimov's Biographical Encyclopedia Of Science & Technology,* 2nd edition. Doubleday and Company 1982.

70. *New Scientist* February 7, 1998 p. 6.

71. *Scientific American Science's Vision: The Mechanics of Sight* 1998, pp. 31–55.

72. *American Family Physician* Vol. 55, no. 6. May 1, 1997, pp. 2219–28.

73. *New Scientist,* June 6, 1998, p. 20.

74. *Discover,* July 1998, p. 86.

75. *The Lancet,* 1999; 353:1585.

76. Tuggy, Michael L. MD. Presented at the 1998 AAFP Scientific Assembly Research competition, San Francisco, September 1998.

77. *Scientific American,* November 1999, pp. 91–97.

78. *Cancer* 1999;86:197–199,324–329.

79. Comroe, Julius H. *Exploring the Heart,* W. W. Norton, 1983, p. 51.

80. *Encyclopedia of Medical History,* McMillian, 1985, p. 34.

81. *JAMA* 1939;113:126–7.

82. *Rh: The Intimate History of a Disease and Its Conquest,* MacMillan 1973.

83. *Proc Soc Exp Biol Med* 1940;43: 223–4.

84. *NY State J Med* 1969;69:2915–2935.

85. *Vox Sanguis* 1984;47:187–90.

86. *Human Biology* 1958;30:14–20.

87. *Vox Sanguis,* op.cit.

88. Race R. R., Sanger R. *Blood Groups in Man.* Oxford, Blackwell Scientific, 1975.

89. *Vox Sanguis.* op. cit.

90. Brochure produced by the Department of Health and Human Services. *Animal Research: The Search for Life-Saving Answers.* Public Health Service, Alcohol, Drug

Abuse, and Mental Health Administration, circa 1991.

91. *The Cambridge World History of Human Disease,* op. cit.

92. *New Scientist* May 3, 1997 p. 22.

93. Comroe, Julius, H. *Exploring the Heart,* W. W. Norton, 1983, pp. 64–7.

94. Comroe, Julius, H. *Exploring the Heart,* W. W. Norton, 1983, p. 67.

95. Editors, The 1997 National Medal of Technology. *Scientific American,* June, 1997.

96. Microsoft (R) Encarta. "X Ray," Microsoft Corporation, Funk and Wagnall's Corporation, 1994.

97. *JAMA* 1999; 282:1041–1046.

98. *New Scientist* May 10, 1997 p. 26.

99. *Scientific American* June 1997 pp. 32–34.

100. *Scientific American* June 1999 pp. 62–69.

101. *Scientific American* June 1999 34.

102. *Mayo Clin Proc* 1997; 72:234.

103. *American Family Physician* Vol. 55, no. 6. May 1, 1997 pp. 2219–28.

104. Wakeley, C. The Changing Pattern of Surgery, *Lancet* November 9, 1957 p. 903.

105. *Medical Tribune* 40 (5):12, 1999.

106. *Reuters,* August 12, 1999.

107. *Clinical Othopaedics and Related Problems* September 1973.

108. Bernard, C. 1865 *Introduction to the Study of Experimental Medicine,* Dover Pub Inc 1957.

109. *Clinical Orthopaedics and Related Research,* no. 211, October 1986.

110. *Animal Models in Orthopedic Research,* An and Friedman (Eds.) CRC Press, 1998, p. 471.

111. *Reuters,* June 10, 1999.

112. *Bone* 1999; 25(no. 2 Supplement): 11S-15S and *Neurochirurgie* 1987; 33:166–8.

113. *The Lancet* 1992; April 11, 339.

114. Altman, Lawrence K. *Who Goes First? The Story of Self-Experimentation in Medicine,* University of California Press 1998, pp. 305–6.

115. Lewin, Roger. *New Scientist,* April 3, 1999, p. 34.

116. *Animal Models in Orthopedic Research,* An and Friedman (Eds.) CRC Press, 1998, p. 370.

117. Greenwald, Robert A. and Dia-mond, Herbert S.(Eds.) *CRC Handbook of Animal Models of Rheumatic Diseases,* Vol. 1, CRC Press 1988.

118. Carlson and Jacobson in Morgan, Douglas W. and Marshall, Lisa A. (Eds.) *In Vivo Models of Inflammation,* Birkhäuser Verlag 1999 p. 38.

119. Carlson and Jacobson in Morgan, Douglas W. and Marshall, Lisa A. (Eds.) *In Vivo Models of Inflammation,* Birkhäuser Verlag 1999 p. 38.

120. *Arthritis Rheum,* 1998;41:1388–1397.

121. *The Cambridge World History Of Human Disease,* Kiple, K. ed. Cambridge 1993.

122. *Animal Models in Orthopedic Research,* An and Friedman (Eds.) CRC Press, 1998, p. 280.

123. *Animal Models in Orthopedic Research,* An and Friedman (Eds.) CRC Press, 1998, p. 6.

124. *Scientific American* May 2000, p. 37.

125. *Nature* 2000;406:143.

126. *Postgraduate Medicine* Vol. 103, no.5, p. 114.

127. *Journal of Organotherapy* Vol. XVI, No. 1, January–February 1932, p. 23.

128. *Am J Clin Nutr* 1997;66:739–40.

129. *Investigative Ophthalmology and Visual Sciences* 1998; 39:344–50.

130. Ibid.

131. *Trans Ophth Soc* 1932; 52:325.

132. *Trans Ophth Soc* 1932; 52:325–52.

133. *Trans Ophth Soc,* 1979; 99:326, *Trans Ophth Soc* 1932; 52:325.

134. *Trans Ophth Soc* 1979;99:325–6, 460.

135. Peterson, Robert A. Affidavit to the Unites States government, The Children's Hospital, Boston. February 28, 1990.

136. *British Journal of Ophthalmology* 1995; 79:585–9.

137. *Annual Review of Neuroscience* 1979; 2:227.

138. *Annual Review of Psychiatry* 1981; 32:47.

139. *Trans Ophth Soc* 1979; 99:432.

140. *J Physiology* 1986;380:453.

141. *Annual Review of Psychology* 1981; 32:477.

142. *British Journal of Ophthamology* 1995; 79:585–9.

143. *Medical Tribune News Service* January 23, 1998.

144. *Medical Tribune* December 1999, Russ Colchamiro.

145. *Medical Tribune* December 1999, Russ Colchamiro.

146. *Science* Vol. 277, September 19, 1997 p. 1805.

147. Nat Genet 1998; 19: 241–247, 257–267.

148. *Reuters Health*, October 30, 1998.

149. *Am J Ophthalmol*, 1953;35:823.

150. *Acta Soc. Ophthalmol Jpn* 1951; 55:219.

151. *Folia Ophthalmol Jpn* 1979;30: 841.

152. *Int Ophthalmol Clin* 1983; 23(3): 129–143.

153. *Ophthalmol Times* 1980;5(12):4–9.

154. *Am J Ophthalmol* 1981;92:292–95.

155. *Lancet*, 1999; 353:1493.

156. *Nature* Vol. 387, May 29, 1997, p. 449.

157. *Lancet* Vol. 349, 1997, 990–993.

158. *Associated Press*, NY, October 9, 1999 1401, EST.

159. *Science*, 1999; 286:2169–2172.

160. *Interagency Coordinating Commit-* tee on the Validation of Alternative Methods press release, June 22, 1999.

161. *British Journal of Ophthalmology* 1999;83:399–402.

162. Azuara-Blanco, Augusto, Pillar, C. T., Dua, Harminder S. *Amniotic membrane transplantation for ocular surface reconstruction*, Department of Ophthalmology, University of Nottingham, England, October 17, 1998.

163. *Mayo Clinic Proc* October 1997, Vol. 72;990–995.

164. *NEJM* October 16, 1997 p. 1132.

165. *Good Medicine*, Winter 1999, pp. 16–17.

166. *Southern Med J* 1994;87:1076–1082.

167. Bruce C. Deighton, Ph.D. Memorial Medical Center Savannah, Ga. *Good Medicine* Vol. IV, no.2.

168. Prof. Dr. Salvatore Rocca Rossetti, surgeon and Professor of Urology at the University of Turin, Italy, in the science program "Delta" on Italian television, March 12, 1986.

169. As quoted in Ruesch, H. *1000 Doctors against Vivisection*, Civis, 1989.

Chapter 6: Pediatrics

1. *Pediatrics* 1999;104:1229–1246.

2. *Teratology* 1974;10:1–7.

3. *Lancet* August 21, 1999.

4. *MMWR* 1999;48:849–57.

5. *Arch Pediatr Adolesc Med* 1997; 151:1082–3, 1096–1103.

6. *NEJM* Vol. 339, no. 5, pp. 313–320.

7. *NEJM* Vol. 339, no. 5, pp. 313–320.

8. *Nature* Vol. 388, p. 434.

9. *New Scientist* August 2, 1997 p. 4.

10. *Lancet* 1999;354:1223–1224,1234–1241.

11. *BMJ* 1993; 307, July 10.

12. *NEJM* 1998;338:147–52.

13. Harte, J., Holdren, C., Schneider, R., and Shirley, C. *Toxins A to Z: A Guide to Everyday Pollution Hazards*, University of California Press, Berkley, 1991.

14. *Epidemiology* 2001:12(2):148 and Reuters Health Newservice, February 16, 2001.

15. *JAMA* 1999; 281:1106–1109.

16. *Epidemiology* 1999;10:661–662, 666–670.

17. *Arch Pediatr Adolesc Med* 1998;152: 127–133.

18. *Pediatrics* 1998;101:E8.

19. *Nat Med* 1998;4:1119–1120, 1144–1151.

20. *Science* 1997; 278:1068–1073.

21. *Nature Medicine*, 1998;4:1119–20 and 1144–51.

22. *New Scientist*, Aug 29,1998, p. 6.

23. *J Am Coll Nutr* 1998;17:379–384.

24. *Arch Gen Psychiatry* 1998; 55:721–727.

25. Association of Birth Defect Children and *Lancet*, 1:1254, 1965.

26. *JAMA*, 1999; 281:1106–1109.

27. *American Journal of Epidemiology*, 1998; 148:173–181.

28. *Am J Epidemiol*, 1998; 148:173–181.

29. *Neurotoxicology and Teratology* Vol. 19, p. 417.

30. *Pediatrics* 1998;101:229–37, 237–41.

31. *Science* 1998; 282;633–634.

32. Anderson, Michael. *The Guardian*, May 5, 1980, p. 13.

33. Dr. K. S. Larsson and colleagues, in a

letter to the journal *The Lancet*, p. 439, August 21, 1982.

34. *Lancet* 1999;354:1441.

35. *Obstetrics and Gynecology* 1998; 92: 187–192.

36. *Journal of Pediatrics* 1999;134:298–303.

37. *MMWR* April, 25, 1997/ Vol. 46/ no. 16.

38. Daniel E. Koshland, Editor-in-Chief of the journal *Science*, Vol 266, p. 1925, 1995.

39. *Nature Medicine*, Vol. 4, no. 8, p. 876, Vol. 3, 60–66 (1997).

40. *Journal of Genetics* 52; 52–67, 1954.

41. *Journal of Nutrition* 123, 27–34, 1993.

42. *Nature Genetics* 13, 275–283, 1996.

43. *Morbidity and Mortality Weekly Review*, August 8, 1997.

44. *The Health Benefits of Smoking Cessation*, Office on Smoking and Health, CDCP.

45. *Pediatrics* 1998;101:229–37, 237–41.

46. *Am J Epidemiol* 1985;121:843–55.

47. *Neurotoxicology and Teratology*, 1993;15:221–260.

48. *Neurotoxicology* 1993;14(1):23–28.

49. *Developmental Brain Research* 1992;69:288–291.

50. *Ann Neurol* 1998;44:665–675.

51. *N Engl J Med* 1999;341:328–335, 364–365.

52. *Drugs* 1996;51(2):226–37.

53. *Pediatric Pulmonology* 1999; 27: 312–17.

54. *A Clinical Guide to Reproductive and Development Toxicology*. CRC Press, Ann Arbor, Michigan 1992.

55. *Acta Anat Nippon* 1969; 74:121–4.

56. *Teratology* 1971;4:427–32.

57. *Lancet* 1998; 351:1197–9.

58. Prof. D. F. Hawkins, Professor of Obstetric Therapeutics & Gynaecology, Hammersmith Hospital, London, in the book *Drugs and Pregnancy: Human Teratogenesis and Related Problems*, publisher Churchill Livingstone, p. 41–49, 1983.

59. *Lancet*, January 26, 1963, p. 222.

60. *Nelsons Textbook of Pediatrics* 1975.

61. Schardein, p. 3–5.

62. Schardein, p. 361–382.

63. *Reuters Health*, July 2, 1999.

64. Schardein, p. 2–3.

65. As quoted in *Bitter Pills* by Stephen Fried Bantam Publishing 1998 p. 274.

66. *Trends in Pharmacological Science*, Vol.8, 1987, p. 133.

67. Schardein, J. L. *Drugs as Teratogens*, CRC Press, 1976.

68. *Environ Health Perspect* 1976;18: 95–96.

69. Shepard, T. H. *Catalog of Teratogenic Agents*, Johns Hopkins University Press, 1980.

70. Schardein, J. L. *Chemically Induced Birth Defects*, Marcel Dekker, Inc., 1993.

71. Schardein, p. 37.

72. Heywood, Ralph. *The Journal of the Royal Society of Medicine*, Vol. 71, pp. 686–689, 1978.

73. Mann, R. D. *Modern Drug Use, An Enquiry on Historical Principles*, MTP 1984.

74. Salen, J.C.W. Animal Models—Principles and Problems. In *Handbook of Laboratory Animal Science* 1994. Svendsen and Hau (eds.) CRC Press 1994 p. 4.

75. *Chest* 2001;119:466–469.

76. Ibid., pp. 83–5.

77. Schardein, pp. 126–46.

78. Schardein, pp. 126–46.

79. Ibid., pp. 416–21.

80. Schardein, pp. 207–11.

81. Ibid., pp. 340–353.

82. Ibid., pp. 340–353.

83. Ibid., pp. 402–09.

84. Ibid., pp. 361–382.

85. Ibid., pp. 426–31.

86. *Reproductive Genetics and the Law*, Mosby-Yearbook 1987, p. 207.

87. *N Engl J Med*, 1998;338:1126–37.

88. Schardein, pp. 90–92.

89. Parke, Dennis. *Animals in Scientific Research: An Effective Substitute for Man?* Ed. Turner. Macmillan, 1983, pp. 7–28.

90. Schardein, pp. 207–11.

91. *NEJM* April 16, 1998 pp. 1128–37.

92. *Ann Neurol* 1999;46:739–746.

93. Association of Birth Defect Children and *Lancet*, 1:1254, 1965.

94. *JAMA*, 1999; 281:1106–1109.

95. *American Journal of Epidemiology*, 1998; 148:173–181.

96. Ibid., pp. 106–17.

97. Ibid., pp. 85–8.

98. Lasagna L. *Drug Use in Pregnancy*, Boston: ADIS Health Science Press 1984.

99. *In vitro Toxicology*, Vol.1, 1987.

100. Yaffe, S. J. *American College of Laboratory Animal Medicine* 1980, p. 13.

101. Lewis, Peter. *Monitoring for Drug Safety*, ed. Inman, 1980, p. 306–313.

102. *Lancet* 1999;354:1659–1660, 1676–1681.

103. Merck Manual.

104. Harrison, M. R., Adzick, N. S., Nakayama, D. K., deLorimier, A. A. Fetal Diaphragmatic Hernia: Pathophysiology, Natural History, and Outcome. *Clinical Obstetrics and Gynecology*, 1986; 29(3).

105. Prof. Dr. Ferdinando de Leo, professor of Pathological and Clinical Surgery at the University of Naples, in an interview with Hans Ruesch for the television station "Teleroma 56" in Rome, May 6, 1986. Translated from Italian.

106. *N Engl J Med* 1998;339:1734–8.

107. *N Engl J Med* 1999;340:1377–82 and 1423–4.

108. *Morbidity and Mortality Weekly Review* August 15, 1997.

109. *BMJ* 1992, Feb 1, 265–266.

110. *BMJ* 1996;313:191–195.

111. *BMJ* 1996;313:195–198.

112. *Int J Epidemiol* 1998;27:238–241.

113. *American Journal of Epidemiology* 1999;149:608–611.

114. *N Engl J Med* 2000;343:262–267.

115. *N Engl J Med* 1998;338:1709–14.

116. National Institute of Child Health and Human Development.

117. Eveld, Edward M. *Kansas City Star,* May 8, 1997.

118. *Transactions of the Ophthalmology Society of Australia,* 1941, Vol. 3, 35–46.

119. Rose and Barker. *Epidemiology for the Uninitiated,* British Medical Journal (publisher), 1986.

120. *Science* 2001; 291:2339.

121. *Principles of Animal Extrapolation.* Calabrese, Edward J. Lewis Publishing 1991 p. 330.

122. *Am J Hum Genetics* 1988;42:585–591.

123. *Nature* 1990;343:183–185.

124. *Science* 1990; 247:566–68.

125. *The EMBO J* 1991;10:3157–65.

126. *Mol and Cell Biol* 1991;11:3786–94.

127. *Proc Natl Acad Sci USA* 1992;89:12155–12159.

128. *Am J Hum Genetics* 1988;42:585–591.

129. *Nature* 1990;343:183–185.

130. *Science* 1990;247:566–568.

131. *The EMBO J* 1991;10:3157–3165.

132. *Mol and Cell Biol* 1991;11:3786–3794.

133. *Proc Natl Acad Sci USA* 1992;89:12155–12159.

134. *NEJM,* July 16, 1998, pp. 194–5.

135. *Am J Med Sci* 1948;215:419–23.

136. *Am J Med* 1960;29:9–17.

137. *Ann NY Acad Sci* 1989;565:222–227.

138. *Proc Natl Acad Sci USA* 1976;73:2033–2037.

139. *Ann Rev of Med* 1992;43:497–521.

140. *Ann Rev of Med* 1992;43:497–521.

141. *NEJM* 1970;282:103–104.

142. *NEJM* 1970;282:103–104.

143. *Br J Haematol* 1981;48:533–43 and *Am J Pediate Hematol Oncol* 1982;4:197–201.

144. *JAMA,* May 12, 1999; 281:no. 18, 1701–1706.

145. *Biochem and Biophys Res Comm* 1987;14:694–700.

146. *New Scientist* December 13, 1997 p. 13.

147. *NEMJ,* Vol. 339, no. 1.

148. *NEJM* 1981; 305:1489–93.

149. *Persp Ped Path* 1976;2:241–78.

150. *Arch Dis Child* 1991;66:698–701.

151. *Pediatrics* 1953;12:549.

152. *Pediatrics* 1953;12:549–563.

153. *Pediatrics* 1951;8:648–656.

154. *Good Medicine,* Cystic Fibrosis: Breakthroughs from Clinical and In-Vitro Research, by Ron Allison, M.D., summer 1993.

155. *Am J of Physiology* 1991;261:491–4.

156. *Proc Nat Acad Sci* 1992;89:5171–75.

157. *Nature* 1990; 347:358–363.

158. *New England Journal of Medicine,* 1998; 339-653-8.

159. *Good Medicine,* Cystic Fibrosis: Breakthroughs from Clinical and In-Vitro Research, by Ron Allison, M.D., summer 1993.

160. *NEJM* 1990;322:291–6.

161. *Science* 1992;43:774–8.

162. *Ped Pulmonology* 1991;6:247.

163. *Drugs* 1992;43(4):431–9.

164. *Proc Nat Acad Sci* 1991;88:10730–4.

165. *Hospital Update* January 1994.

166. Lee, T. *Gene Future,* Plenum Press 1993 p. 177.

167. *Science* 1990;249:537–540.

168. *Proc Natl Acad Sci USA* 1991;88: 10730–10734.

169. *Exp Lung Res* 1990;16:661–670.

170. *Cell* 192;68:143–155.

171. *ATLA* 1998;26–27.

172. *Medical Tribune* 40(8):21, 1999.

173. *Science* 1999;286:388–389,544–548.

174. *Proc Natl Acad Sci U.S.A.* 2000;97: 11614–11619.

175. *J Pediat* 1952;40:767–771.

176. *New Scientist* December 7, 1991: pp. 30–34.

177. *NEJM* 1992;326:812–815.

178. *Pediatrics* 1959;24:739–745.

179. *Am Rev Resp Dis* 1963;88:199–204.

180. *Pediatrics* 1961;27:589–596.

181. *Proc Natl Acad Sci USA* 1990;87: 9188–9192.

182. Salen, J.C.W. Animal Models—Principles and Problems. In *Handbook of Laboratory Animal Science* 1994, Svendsen and Hau (eds.) CRC Press 1994 p. 5.

183. *JAMA* 1992;267:14, 1947–9.

184. *N Engl J Med* 1990;322:1189–94.

185. *European Respiratory Review* 1995 5:29 p. 241.

186. *European Respiratory Review* 1995 5:29 p. 184.

187. *Nature Genetics* 1997;17:370–1, 411–12.

188. *Nature* Vol. 387, pp. 80–83.

189. *Science* 1997; 278:1315–18.

190. *Pediatrics* 1999;103:546–550.

191. *Lancet* 1998;351:383–4, 394–98, 415.

192. *Science* Vol. 279, March 20, 1998, pp. 1870–1, 1950–4.

193. *Neurology* 1998;50:114–120.

194. *JAMA* 1999;282:2125–2130,2167–2168.

195. *Horizons in Medicine,* no. 6, Blackwell Science Ltd. p. 28.

196. *Ann Neurol* 1998;43:143–147.

197. *Nature Genetics* Vol. 15, 1997, pp. 87–90.

198. *Nature* Vol. 391, January 8, 1998 p. 184.

199. *Nat Gen* 1999; 23:127–128,185–188.

200. *Scientific American* August 1977, pp. 40–47.

201. *NEJM* November 20, 1997p. 1548.

202. *Scientific American* December 1977 pp. 68–73.

203. *Nature Genetics* 1997;18:14–15.

204. *Nature* Vol. 392, April 30, 1998 p. 923.

205. *Pediatrics* 1998;101:306–308.

206. *Blood* 1999; 94:3678–3682.

207. *Medscape,* December 6, 1999.

208. *N Engl J Med* 1999;341:1180–1189,1227–1229.

209. Maestripieri, Dario and Carroll, K. A. *Child abuse and neglect: Usefulness of the animal data, Psychological Bulletin* 123(May):211.

210. *Biol Neonate* 1992; 61:131–136.

211. *Am J Obstet Gynec* 1991;165(2): 438–442.

212. *J Mol Cell Cardiol* 1992; 24:1409–1421.

213. *American Physiology Society* 1987 R306-R313.

214. *Neuropharmacology* 1991;30(1): 53–58.

215. *Neuropharmacology* 1991;40:429–432.

216. *Neuropharmacology* 1991; 30(1): 53–58.

217. *Neuropharmacology* 1991;40:429–432.

218. Hill, R. B. and Anderson, R. E. *The Autopsy—Medical Practice and Public Policy,* Butterworth 1988.

219. *Lancet* 1999; 354:1223–1224,1234–1241.

220. *American Journal of Obstetrics and Gynecology* 1998; 179:203–209.

Chapter 7: Diseases of the Brain

1. Hill, R. B. and Anderson, R. E. *The Autopsy—Medical Practice and Public Policy,* Butterworth 1988.

2. Hill, R. B. and Anderson, R. E. *The Autopsy—Medical Practice and Public Policy,* Butterworth 1988.

3. *Reuters Health,* November 23, 2000.

4. *Lancet* June 16, 1883.

5. *Lancet* August 16, 1958 p. 365.

6. *Neurology* 1981; 31:600–02.

7. As quoted in Bayly, B. *Clinical Medical Discoveries.* National Antivivisection Society, London, 1961 p. 27.

8. *Neuroreport* Vol.9, p. 1013.

9. *Motor Control* 2001 July; 5(3):222–30.

10. *J Am Parapl Soc* 11;23–25, 1988.

11. *New England Journal of Medicine* 2001;344:556–563,602–603.

12. *Arch Neurol* 1968;19:472–86, *Neurology* 1960;10:499–505, *Am J Physiol* 1955;183:19–22, *Surg Gynecol Obstet* 1960;110:27–32.

13. Immunology and Cell Biology 1998; 76:47–54.

14. *Mayo Clinic Health Letter,* November 1995, updated January 1998.

15. *NEJM* 1998;338:278–285,323–25.

16. *Nat Med* 2000;6:1098–1100, 1167–1182.

17. *Neuroscience News,* October 1, 2000.

18. *Immunological Reviews* 1999;169: 68.

19. Ann Neurol, 2000;47:691–693,707–717.

20. *Immunologic Research* 1998;17:217–227.

21. *Scientific American* Vol. 268, 1993, pp. 81–82.

22. *J Mol Med* 1997;75:187–97.

23. *Experimental and Molecular Medicine* 1999;31:115–121.

24. *Scientific American* Vol. 268, 1993, pp. 81–82.

25. *Mayo Clinical Proc* 1997; 72: 663–78.

26. *Immunology and Cell Biology* 1998; 76:47–54.

27. *J Neurol Neurosurg Psychiatry* 1998;64:7300–735.

28. *J of Neurology, Neurosurgery and Psychiatry* 2000;69:25–28.

29. *Mayo Clinic Health Letter,* November 1995, updated January 1998.

30. *Nature Medicine* 1997;3:1394–97.

31. *Ann Neurol* 1998;44:629–634.

32. *Neurology* 1995;45:1268–76.

33. *Immunology and Cell Biology* 1998; 76:65–73.

34. *J Neurol Neurosurg Psychiatry* 2000; 68:89–92.

35. *Nature Medicine* 2000;6:1167–82.

36. *Immunological Reviews* 1999;169:5–10.

37. *Mayo Clinic Health Letter,* November 1995, updated January 1998.

38. *Lancet* 1987;1:893–5 and *Neurology* 1987;37:1097–1102.

39. *Am J Hum Genet* 1998;62:633–40.

40. *Cell* Vol. 20, p. 589.

41. *Laboratory Animal Science* 1999;49: 480–7.

42. *Pharmaceut Res* 1998;15:386–98.

43. *N Engl J Med* 1999;340:1970–9.

44. *N Engl J Med* 1999;340:1970–9.

45. *Proc Natl Acad Sci USA* 1998;95: 000–000.

46. *Ann Neurol* 1998;00:000–000.

47. *Science* Vol. 280, June 5, 1998 p. 1524–5.

48. *Nature Medicine* Vol. 4, no. 7, p. 827–31.

49. *JAMA* 1997; 277: 813–17.

50. *N Engl J Med* 1999;340:1970–9.

51. *Scientific American,* December 2000, p. 78.

52. *Arch Neurol* 1974;30:113–21, Hollister, L. *Chemical Psychoses,* Charles C. Thomas 1968, and *Pharmacol Review* 1966;18:965–996.

53. See Bartus et al, Usdin, E. (Ed.) *Psychopharmacology: A Third Generation of Progress* Raven Press 1987 p. 219–32.

54. *Brain* 1976;99:459–96, *Lancet* 1976; ii:1403, *Lancet* 1977; ii:189.

55. *Integrat Psychiatry* 1986;4:74–75.

56. Tanzi, Rudolph. *Nature Medicine* 1998;4:1127–8.

57. *Nature* 2000;405:689–95.

58. *Mol Cell Neurosci* 1999;14:419–27.

59. *Nature* 1999;398:466–467, 513–517.

60. *Nature* Vol. 387, May 15, 1997.

61. Mconner and Tuszynski in Emerich, D. F., Dean III, R. L., and Sanberg, P. R. (Eds) *Central Nervous System Diseases: Innovative Animal Models from Lab to Clinic* Humana Press 2000, p. 66.

62. *Scientific American,* December 2000, p. 81.

63. *Science* 1999;286:735–41, *Molecular and Cellular Neuroscience* as quoted in *Nature* 1999;402:471, *Nature* 1999;402: 533–7, and 537–40.

64. *Nature* 1999;402:471–472,533–540.

65. *Nature Biotechnology* 2000;18:125–6.

66. *Nature* 2000: 407:34–35,48–54.

67. *Arch Neurol* 1998;55:964–968.

68. *Nature Genetics* 1998;18:211–12.

69. *Scientific American,* December 2000, p. 82.

70. *Reuter's Health,* August 18, 2000.

71. *Science* 2000;408:2302–5.

72. *Am J Pathol* 1999;155:1163–1172.

73. *JAMA* 1998;280:614–18.

74. *JAMA* 1998;280:19–22.

75. *JAMA* 1998;280:652–3.

76. *JAMA* 2000;283:1571–77.

77. *N Engl J Med* 1999;340:1970–9.

78. *Scientific American,* December 2000, p. 83.

79. *Nature Genetics* 1998;18:211–12.

80. *Nat Genet* 1998;19:314–316,321–322,357–360.

81. *Ann Neurol* 2000;47:361–368.

82. *Nature Medicine* Vol.4, no. 7, pp. 832–4.

83. *Arch Gen Psychiatry* 1999;56:981–987,991–992.

84. *Toronto /MedscapeWire,* American Academy of Neurology 51st Annual Meeting, April 17–24, April 23, 1999.

85. *Nature Medicine* 2000;6:20.

86. *Proc Natl Acad Sci USA* 1999;96: 14079– 84.

87. *Neurology* 1998;51:1009–1013.

88. *Neurology* 2001;56:950–956.

89. *J Neural Transm* Suppl. 1997;49: 33–42. Review.

90. *Nature* 1999;400:173–7.

91. *Nature* 1999;400:116–17.

92. *N Engl J Med* 1999;341:1694–5.

93. *Science* 2000;408:915.

94. *Science* 2000;404:915–16, 975–85.

95. *Ann Neurol* 1997;42:776–82.

96. *Scientific American,* December 2000, p. 83.

97. *American Family Physician* Vol. 56, no. 4, p. 1166.

98. *Archives of Neurology* 1998;55: 1449.

99. *BMJ* 1999;519:807–811.

100. *Reuter's Health,* October 20, 1999.

101. *Nature* 1999;401:376–9.

102. *Ann Neurol* 1997;42:776–82.

103. *Nature* 1999;401:376–79.

104. *New England Journal of Medicine* 1998;339:1044–53.

105. *Science* 2001;291:567–9.

106. *Science* 2001;291:567–9. The letter was published in *Parkinsonism and Related Disorders* in March 2000.

107. Scallert, T. and Tillerson, J. L. in Emerich, D. F., Dean III, R. L., and Sanberg, P. R. (Eds.) *Central Nervous System Diseases: Innovative Animal Models from Lab to Clinic* Humana Press 2000, p. 131.

108. Kau and Creese in Emerich, D. F., Dean III, R. L., and Sanberg, P. R. (Eds.) *Central Nervous System Diseases: Innovative Animal Models from Lab to Clinic,* Humana Press 2000.

109. *Nuc Med Biol* 1998;25:721–8.

110. *Current Opinion in Neurology* 1996;9:303–7.

111. *Review in the Neurosciences* 1998; 9:71–90.

112. Gerlach, M. and Riederer, P. *Journal of Neural Transmission* 1996;103:987–1041.

113. *Journal of the Neurological Sciences* 1973;20:415–55.

114. *Reuter's Health* October 18, 2000.

115. *Neurology* 1998;51:1057–1062.

116. *British Medical Journal* 1998;317: 1033.

117. *Reuter's Health,* November 17, 1999.

118. Parkinson's Disease: Medical Research Council (UK) Annual Report, 1983–1984; and Alzheimer's Disease: WK Summers et al, *Biological Psychiatry,* 1981, Vol.16, 145–153.

119. *Science* Vol. 276, June 27, 1997.

120. *Science* Vol. 277, July 18, 1997 p. 387.

121. *Nature Genetics* 1998;18:106–108.

122. *Nature* 1997; 388:839–40.

123. *Nature* 393 p. 702.

124. *Science* Vol. 276, June 27, 1997.

125. *Science* Vol. 277, July 18, 1997 p. 387.

126. *Nature* Vol. 329, April 9, 1998, p. 605.

127. *Nature Genetics* 1998;18:262–265.

128. *Lancet*1999;354:1658–1659,1665–1669.

129. *Nat Neurosci* 1999;2:1047–1048,1137–1140.

130. *Epilepsia* 1997;38:957.

131. *Current Opinion in Neurology* 1998;11:123–27.

132. *Science* Vol. 277, August 8, 1997 p. 805–808.

133. *Epilepsia* 1999;40:811–21.

134. Sperber, E. F. et al ch. 8 p. 161–169 in *Jasper's Basic Mechanisms of the Epilepsies, Third Edition: Advances in Neurology* Vol. 79, edited by A. V. Delgado-Escuetta, W. A. Wilson, R. W. Olsen and R. J. Porter. Lippincott, Williams & Wilkins 1999.

135. *Epilepsia* 1999;40(Suppl. 1):S51–58.

136. Jeffrey L Noebels. Chapter 12 p. 227–238 in *Jasper's Basic Mechanisms of the Epilepsies, Third Edition: Advances in Neurology* Vol. 79, edited by A. V. Delgado-Escuetta, W. A. Wilson, R. W. Ol-

sen and R. J. Porter. Lippincott Williams & Wilkins 1999.

137. *Epilepsia* 1999; 40(Suppl. 3):17–22 and Jeffrey L Noebels. Chapter 12 p. 227–238 in *Jasper's Basic Mechanisms of the Epilepsies, Third Edition: Advances in Neurology* Vol. 79, edited by A. V. Delgado-Escuetta, W. A. Wilson, R. W. Olsen and R. J. Porter. Lippincott, Williams & Wilkins 1999.

138. Seyfried et al. Chapter 15 p. 279–90 in *Jasper's Basic Mechanisms of the Epilepsies, Third Edition: Advances in Neurology* Vol. 79, edited by AV Delgado-Escuetta, W. A. Wilson, R. W. Olsen and R. J. Porter. Lippincott Williams & Wilkins 1999

139. *Munch Med Wochenschr* 1912;59: 1907–9

140. Coulston and Shubick (Eds.) *Human Epidemiology and Animal Labroratory Correlations in Chemical Carcinogenesis*, Ablex Publishing, 1980, p. 251

141. *Epilepsia* 1996;37(suppl. 6):S1–S3

142. *Epilepsia* 1996;37(suppl. 6):S1–S3

143. *J Neurol Neurosurg Psychiatry* 1999;67:707–708, 716–722

144. *Progress in Neurobiology* 1997;53: 239–58

145. *Progress in Neurobiology* 1997;53: 239–58

146. *Epilepsia* 1997; 38: 1262–4

147. *JAMA* 1938; 111: 1068–73

148. Parham and Bruinvels (Eds.) *Discoveries In Pharmacology. Volume I Psycho- and Neuro-pharmacology* Elsevier 1983 p. 447–77

149. Parham and Bruinvels (Eds.) *Discoveries In Pharmacology. Volume I Psycho- and Neuro-pharmacology* Elsevier 1983 p. 454

150. *Ann Neurology* 1996;40:908–11

151. *Journal of Neurochemistry* 1988, Vol.50, 225–9

152. Dr. Med. Bernhard Rambeck, since 1975 director of the Biochemistry Department of the Society for Epilepsy Research in Bielefeld-Bethel, West Germany. From his speech at International Symposium of April 25,1987, Zurich

153. *Nature Medicine* 1998;4:1173–6

154. Feb 20, 2001 Reuters Health and *Journal of Combinatorial Optimization.* Issue unknown

155. Isaac R. J., Armat V. C. *Madness in the Streets: How Psychiatry and the Law Abandoned the Mentally Ill.* New York:

Free Press, 1990, and Kramer, P.D., *Listening to Prozac,* New York: Penguin 1993

156. Stone, Trevor and Darlington, Gail. *Pills, Potions, Poisons,* Oxford University Press, Oxford, England, 2000

157. *Chlorpromazine in Psychiatry: A Study in Therapeutic Innovation.* Cambridge, MIT Press, 1974 and Reines BP, *On the Locus of Medical Discovery. The Journal of Medicine and Philosophy* and Caldwell, A., *Origins of Psychopharmacology: From CPZ to LSD.* Springfield, Charles C. Thomas, 1970

158. Parham and Bruinvels. *Discoveries In Pharmacology,* Elsevier 1983 p. 164–67

159. *Psychiatric Times* September 1987

160. *Psychiatric Times* September 1987

161. Schwiez *Med Wochenschn* 1957; 87:1135–1140

162. Sulser F, Mishra R. The Discovery of Tricyclic Antidepressants and Their Mode of Action, in Parham MJ, Bruinvels J: *Discoveries in Pharmacology*: Vol. 1, Psycho- and Neuro-Pharmacology. New York, Elsevier, 1983)

163. Parham and Bruinvels (Eds). *Discoveries In Pharmacology. Volume I Psycho- and Neuro-pharmacology* Elsevier 1983, p. 270

164. *Chemotherapy in Psychiatry.* Cambridge, Mass. Harvard University Press. 1977

165. Parham and Bruinvels. *Discoveries In Pharmacology. Volume I Psycho- and Neuro-pharmacology* Elsevier 1983. p. 233–47

166. Graef, F. G. In Simon, P et al, *Selected Models of Anxiety, Depression, and Psychosis* Vol. 1 1988, p. 115

167. *J Clin Psychiatry* 44: 5 [Sec. 2] 40–48, 1983

168. Lehman H., Kline, N. S. Alleviation of anxiety—the benzodiazepine saga, in Parnham NJ, Bruinvels, J.: *Discoveries in Pharmacology*: Vol.1, Psycho- and Neuro-Pharmacology. New York, Elsevier, 1983

169. *J Clin Psychiatry* 44: 5 [sec. 2] 40–48, 1983

170. *Prog in Neuropsychopharmacol Biol Psychiat* 1983; 7: 227–8

171. *Br Nat Formulary,* no. 26, 1993

172. *Br J Clin Pharm,* 1983, Vol.15, 291S-3

173. *Drug Metab & Dis,* 1991, Vol.19, 841–3

174. *J Clin Psychiatry* 44: 5 [sec. 2] 40–48, 1983

175. Weil-Malherbe, H. and Sarza, S. I. *The Biochemistry of Functional and Experimental Psychosis*, C. C. Thomas Publisher, 1971, p. 57–76

176. *Arch Gen Psych* 1972;26:472–78

177. *Brit J Psychiatry* 1967;113:1407–11

178. *Lancet,* 1968; II:805–808

179. *Adv Biochem Psychopharmacol* 1974;11:387–98

180. *Lancet,* 1986;2:1049–52

181. *J Neurochem* 1966;13:1545–48

182. *J Clin Psych* 1986;47:23–35

183. *Lancet* 1963;1:79–81 and 527–528

184. *Lancet* 1967; 2:1178–80

185. *Lancet,* 1968;1: 531–2

186. *Am J Psychiat* 1963;120:274–5

187. *Drugs Exptl Clin Res* 1977;1:239–42

188. *Neuropharmacology* 986;25;799–802

189. *J Clin Psychiatry* 44: 5 [sec. 2] 40–48, 1983

190. *Nature* 1992;360:160–63

191. *Biological Psychiatry,* October 15, 2000

192. *Prog Neuro-Psychopharmacol & Biol Psychiat* 1983;7:227–228

193. *American Psychiatric Association DSM-IV.* American Psychiatric Association 1994

194. *American Journal of Psychiatry,* 2000;157:1619–1628, Medscape Wire, October 4, 2000

195. *Molecular Psychiatry* 1997;2:463–471

196. *Nature Genetics* 10: 41–46

197. *Am J of Med Genet* 2000;96:56–60.

198. *Scientific American,* September 1998, pp. 67–71

199. *Am J of Psych* 1999;156:252–257

200. *The Medical Letter* Vol.40, Issue 1020

201. Gallup, G. G., and Suarez, S. D. *Psychol Rec* 30:211–218

202. Kelly, J. A. *Am Psych* 1986;41:839–41

203. *Am Psych* Vol.51, no.11, November 1996 p. 1167–1179

204. *New Scientist* December 18, 1999 p. 13

205. Davis, J. M. *Antipsychotic Drugs,* in Kaplan, H. I., Sadock, B. J., 9 eds. *Comprehensive Textbook of Psychiatry,* Fourth Edition. Baltimore, Williams and Wilkins, 1985

Chapter 8: Beyond the Animal Model

1. *Arch Pathol Lab Med,* Vol. 120, August 1996.

2. *JAMA* 1998; 280:1273–4 and *JAMA* 1998: 280:1245–8.

3. *Chest* 2001;119: 530–536.

4. *JAMA* 2001; 285:540–544.

5. *Science* 1998; 282:396–9.

6. *Science* 1999; 285:998–1002.

7. *Science* 1999;286: 487–491 and Westport Newsroom 203 319 2700.

8. *Science* 1998; 282:396–9.

9. *ATLA* 1998; 26:27.

10. *ATLA* 1998; 26:27.

11. *Comparative Medicine* 2000;50:10–11.

12. van Zutphen, L.F.M. PhD of the department of Laboratory Animal Science, Utrecht University The Netherlands in Comparative Medicine 2001;51:110

13. VandeBerg, John L. in Bennett, Abee, and Henrickson (Eds.) *Nonhuman Primates in Biomedical Research, Biology and Management.* Academic Press 1995, p. 140.

14. *Nature* 1999; 400:309–10

15. *New England Journal of Medicine* 2000; 342:42–49.

16. *Medical Tribune* 40(8):16, 1999.

17. *Scientific American Science's Vision: The Mechanics of Sight* 1998, p. 31–55.

18. *JAMA* 1998; 280:1510–1516.

19. Scientific Fraud and Misconduct and the Federal Response: Hearing before the Human Resources and Intergovernmental Relations Subcommittee of the Committee on Government Operations, April 11, 1998, p. 108

20. Bragg, Melvyn. *On Giants' Shoulders.* John Wiley and Sons 1998, p. 31.

21. National Cancer Advisory Board, Minutes: Report of the Clinical Investigators Task Force, Bethesda, MD September 20–21, 1993, *Journal of Clinical Oncology* Vol. 1996; 14, no. 2

Postscript

1. Russell, Bertrand. *The Autobiography of Bertrand Russell.* New York, 1968: Atlantic Monthly Press.

Index